Clinical Mitochondrial Medicine

Clinical Mitochondrial Medicine

Edited by

Patrick F. Chinnery
University of Cambridge

Michael J. Keogh
University of Cambridge

CAMBRIDGE
UNIVERSITY PRESS

Shaftesbury Road, Cambridge CB2 8EA, United Kingdom

One Liberty Plaza, 20th Floor, New York, NY 10006, USA

477 Williamstown Road, Port Melbourne, VIC 3207, Australia

314–321, 3rd Floor, Plot 3, Splendor Forum, Jasola District Centre, New Delhi – 110025, India

103 Penang Road, #05–06/07, Visioncrest Commercial, Singapore 238467

Cambridge University Press is part of Cambridge University Press & Assessment, a department of the University of Cambridge.

We share the University's mission to contribute to society through the pursuit of education, learning and research at the highest international levels of excellence.

www.cambridge.org
Information on this title: www.cambridge.org/9780521132985

DOI: 10.1017/9781139192460

First published 2018 (version 4, October 2024)

Printed in Great Britain by Ashford Colour Press Ltd., January 2024

A catalogue record for this publication is available from the British Library

Library of Congress Cataloging-in-Publication data
Names: Chinnery, Patrick F., editor. | Keogh, Michael (Michael J.), editor.
Title: Clinical mitochondrial medicine / edited by Patrick Chinnery, Michael Keogh.
Description: Cambridge ; New York, NY : Cambridge University Press, 2018. | Includes
    bibliographical references and index.
Identifiers: LCCN 2017055396 | ISBN 9780521132985 (pbk. : alk. paper)
Subjects: | MESH: Mitochondrial Diseases
Classification: LCC RB155 | NLM WD 200.5.M6 | DDC 616/.042–dc23
LC record available at https://lccn.loc.gov/2017055396

ISBN 978-0-521-13298-5 Paperback

Contents

*Color plates are to be found between
pp. 114 and 115*

Contributors

Laurence A. Bindoff MD, PhD
Department of Neurology, Haukeland University Hospital and Department of Clinical Medicine (K1) University of Bergen, Norway

Valerio Carelli MD, PhD
IRCCS Institute of Neurological Sciences of Bologna, Bologna, Italy

Patrick F. Chinnery FMedSci, FRCP, PhD
MRC Mitochondrial Biology Unit & Department of Clinical Neurosciences, University of Cambridge, United Kingdom

Thomas M. F. Connor MRCP, PhD
Imperial College London, London, United Kingdom

Irenaeus F. M. de Coo MD, PhD
Department of Neurology, Erasmus Medical Center, Rotterdam, the Netherlands

Marni J. Falk MD
Division of Human Genetics, The Children's Hospital of Philadelphia and University of Pennsylvania Perelman School of Medicine, Philadelphia, Pennsylvania, United States

Grainne S. Gorman MRCP, PhD
Wellcome Centre for Mitochondrial Research and Institute of Neuroscience, Newcastle University, United Kingdom

Michael G. Hanna MD, FRCP
MRC Centre for Neuromuscular Diseases and Department of Molecular Neuroscience, University College London, United Kingdom

Alison J. Heggie MRCP
Institute of Cellular Medicine, Newcastle University, United Kingdom

Rita Horvath MD, PhD
Wellcome Centre for Mitochondrial Research, Institute of Genetic Medicine, Newcastle University, United Kingdom

Jessie M. Hulst MD, PhD
Department of Paediatric Gastroenterology, Erasmus Medical Center, Sophia Children's Hospital, Rotterdam, the Netherlands

Michael J. Keogh BMBS (Hons), MRCP
Department of Clinical Neurosciences, University of Cambridge, United Kingdom

Peter Kullar MRCS, DOHNS
ENT Department, Freeman Hospital, Newcastle-upon-Tyne, United Kingdom

Patrick H. Maxwell FMedSci, FRCP, D Phil
School of Clinical Medicine, University of Cambridge, United Kingdom

Philip G. Morgan MD
Seattle Children's Research Institute, Seattle, United States

Chiara La Morgia MD, PhD
IRCCS Institute of Neurological Sciences of Bologna, Bologna, Italy

Colleen Clarke Muraresku MS
Division of Human Genetics, The Children's Hospital of Philadelphia, Philadelphia, Pennsylvania, United States

Robert D. S. Pitceathly MRCP, PhD
MRC Centre for Neuromuscular Diseases and Department of Molecular

Neuroscience, University College London,
United Kingdom

Shamima Rahman FRCP, PhD
UCL Great Ormond Street Institute
of Child Health, London,
United Kingdom

Mary M. Reilly FRCP, MD
MRC Centre for Neuromuscular Diseases
and Department of Molecular
Neuroscience, University College London,
United Kingdom

Andrew M. Schaefer MRCP
Wellcome Centre for Mitochondrial
Research and Institute of Neuroscience,
Newcastle University, United Kingdom

Margaret M. Sedensky MD
Seattle Children's Research Institute,
Seattle, United States

Hannah E. Steele MBBS (Hons), MRCP
Institute of Genetic Medicine, Newcastle
University, United Kingdom

Robert W. Taylor PhD, DSc, FRCPath
Wellcome Centre for Mitochondrial
Research and Institute of Neuroscience,
Newcastle University, United Kingdom

**David R. Thorburn PhD, FHGSA, FFSc
(RCPA)**
Murdoch Children's Research Institute,
Victoria 3052, Australia

Carlo Viscomi PhD
MRC Mitochondrial Biology Unit,
Cambridge, United Kingdom

Mark Walker FRCP, MD
Institute of Cellular Medicine, Newcastle
University, United Kingdom

Patrick Yu-Wai-Man FRCOphth, PhD
Department of Clinical Neurosciences,
University of Cambridge, United
Kingdom

Massimo Zeviani MD
MRC Mitochondrial Biology Unit,
Cambridge, United Kingdom

Preface

Although mitochondrial diseases affect fewer than 1 in 5,000 adults and children, they enter the differential diagnosis of many common medical disorders. It is therefore important that all practicing clinicians have an understanding of the clinical presentation of mitochondrial disorders, if only to recognize the possibility and to refer on to a specialist for further evaluation. To the uninitiated, mitochondrial diseases are complex and confusing, with vague symptoms accompanying a myriad of clinical signs. They are difficult to diagnose with confidence, and even more difficult to exclude. Keeping abreast of the rapidly expanding online literature seems to have turned into a full-time job, so how can the generalist even begin to approach the problem? *Clinical Mitochondrial Medicine* aims to provide a practical and comprehensive guide to clinicians who may meet patients with suspected mitochondrial disease from time to time in their daily practice.

This book has been written by an international group of authors, most of whom see patients with mitochondrial disease on a regular basis. They have deliberately focused on key facts required to recognize, diagnose and manage patients and families with mitochondrial disease. Discussions of the disease mechanisms have been kept to a bare minimum, and are included only when it helps the reader to understand the clinical presentation, approach the investigation and develop a management strategy.

Although this book can be used as a reference text, we have taken great steps to ensure limited overlap between the chapters. This means that both the text and the figures in each chapter are relevant throughout this book, and it is recommended that new readers cast their eye over the whole volume before using it in practice. Inevitably, with a rapidly moving field, some aspects will become out of date. However, we believe that the core principles – which form the basis of this book – will stand the test of time.

It has been enormously exciting to be involved in the development of mitochondrial medicine. Inevitably for a young field, some of the chapters include opinion based on anecdote and clinical experience, rather than evidence-based practice. However, many of those experienced views are now being endorsed through large-scale natural history studies of clinical cohorts assembled through international consortia. Unfortunately this approach has had little impact on treatment to date, but the first randomized controlled clinical trials have now been carried out in rare mitochondrial disorders, and more than 10 are currently in the pipeline (www.clinicaltrials.gov) led by clinicians, academics and the pharmaceutical industry. It is therefore an stimulating time to be involved in the care of patients with mitochondrial disease, as we enter a phase of evidence-based medicine leading to new management approaches aimed at preventing and treating these hitherto incurable diseases.

Patrick F. Chinnery and Michael J. Keogh
Cambridge, Jan 2018

Abbreviations

AAV	Adeno-associated virus
ABR	Auditory brainstem response
ACE	Angiotensin-converting enzyme
ACR	Albumin-creatinine ratio
ACTH	Adrenocorticotropic hormone
ADAOC	Autosomal dominant optic atrophy and early-onset cataracts
ADP	Air displacement plethysmography
adPEO	Autosomal-dominant PEO
ADTKD	Autosomal-dominant tubulointerstitial kidney disease
AGK	Acylglycerol kinase
AKI	Acute kidney injury
ANS	Ataxia Neuropathy Spectrum
ARB	Angiotensin receptor blocker
ARHL	Age-related hearing loss
arPEO	Autosomal-recessive PEO
ATP	Adenosine triphosphate
AQP4	Aquaporin-4 antibody
AV	Atrioventricular
BBGD	Biotin-responsive basal ganglia disease
BBVL	Brown-Vialetto-Van Laere syndrome
BIA	Bioelectrical impedance analysis
BiPAP	Bi-level positive airway pressure
BMI	Body mass index
CCDD	Congenital cranial dysinnervation disorder
CFEOM	Congenital fibrosis of the extraocular muscles
CIDP	Chronic inflammatory demyelinating polyradiculoneuropathy
CIPO	Chronic intestinal pseudo obstruction
CK	Creatine kinase
CKD	Chronic kidney disease
CMR	Cardiac magnetic resonance imaging
CMT	Charcot-Marie-Tooth
CMV	Cytomegalovirus
CNS	Central nervous system
CoQ_{10}	Coenzyme Q_{10}
COX	Cytochrome c oxidase
CPAP	Continuous positive airway pressure
CPEO	Chronic progressive external ophthalmoplegia
CPT1&2	Carnitine palmitoyltransferase 1 and 2
CSF	Cerebrospinal fluid
CT	Computer tomography
CVD	Coronary vascular disease
CVS	Chorionic villi sampling

DAB	Diaminobenzidine tetrahydrochloride
DCM	Dilated cardiomyopathy
DEXA	Dual-energy x-ray absorptiometry
DOA	Dominant optic atrophy
DPOAE	Distortion product otoacoustic emission
DPP41	Dipeptidyl peptidase 4 inhibitor
DRP1	Dynamin-related protein 1
ECFV	Effective circulating fluid volume
ECG	Electrocardiogram
ECHO	Echocardiogram
EEG	Electroencephalograph
EMA	European Medicines Agency
EMG	Electromyography
EN	Enteral nutrition
EPS	Electrophysiological studies
ERG	Electroretinography
ESRD	End-stage renal disease
FAD	Flavin adenine dinucleotide
FAO	Fatty acid oxidation
FBC	Full blood count
FEES	Fiber-optic endoscopic evaluation of swallowing
FEV	Forced expiratory volume
FMN	Flavin mononucleotide
fMRI	Functional MRI
FRDA	Friedreich ataxia
FSGS	Focal segmental glomerulosclerosis
FSH	Follicle-stimulating hormone
FVC	Forced vital capacity
GDM	Gestational diabetes mellitus
GFR	Glomerular filtration rate
GH	Growth hormone
GMFM	Gross motor function measure
GRACILE	Growth retardation, aminoaciduria, cholestasis, iron overload, lactic acidosis, and early death
HCM	Hypertrophic cardiomyopathy
H&E	Hematoxylin and eosin
HMSN-6	Hereditary motor and sensory neuropathy type 6
HSCT	Hematopoietic stem cell transplantation
HMSN	Hereditary motor and sensory neuropathy
HSP-7	Hereditary spastic paraplegia type 7
ICD	Implantable cardioverter defibrillator
IOSCA	Infantile onset spino-cerebellar ataxia
IMS	Inter-membrane space
ISCEV	International Society for Clinical Electrophysiology of Vision
IVF	In vitro fertilization
JVP	Jugular venous pressure
KSS	Kearns-Sayre syndrome

LBSL	Leukoencephalopathy with brain stem and spinal cord involvement with lactate elevation
LBTL	Leukoencephalopathy with thalamus and brainstem involvement and high lactate
LCHADD	Long-chain hydroxyacyl-CoA dehydrogenase deficiency
LGE	Late gadolinium enhancement
LH	Luteinizing hormone
LHON	Leber hereditary optic neuropathy
LR	Lactated Ringer's
LV	Left ventricular
LVH	Left ventricular hypertrophy
LVNC	Left ventricular non compaction
LVOTO	Left ventricular outflow obstruction
MADD	Multiple acyl-CoA dehydrogenase deficiency
MD	Mitochondrial disorder
MDDS	Mitochondrial DNA depletion syndromes
MEGDEL	3-methylglutaconic aciduria, deafness, encephalopathy and Leigh-like disease
MELAS	Mitochondrial encephalomyopathy, lactic acidosis and stroke-like episodes
MEMSA	Myoclonic epilepsy with myopathy and sensory ataxia
MERRF	Myoclonic epilepsy with ragged red fibers
MH	Malignant hyperthermia
MIDD	Maternally inherited diabetes and deafness
MILS	Maternally inherited Leigh syndrome
MIMyCa	Maternally inherited adult-onset myopathy and cardiomyopathy
MIRAS	Mitochondrial recessive ataxia syndrome
MNGIE	Mitochondrial neurogastrointestinal encephalomyopathy
MRI	Magnetic resonance imaging
MRS	Magnetic resonance spectroscopy
MS	Multiple sclerosis
mtDNA	Mitochondrial DNA
nDNA	Nuclear DNA
NADH	Nicotinamide adenine dinucleotide (reduced form)
NAG	N-Acetyl-β-D-glucosaminidase
NARP	Neuropathy ataxia and retinitis pigmentosa
NBT	Nitroblue tetrazolium
NCS	Nerve conduction studies
NGAL	Neutrophil gelatinase-associated lipcalin
NGS	Next-generation DNA sequencing
NICE	National Institute for Health and Care Excellence
NMO	Neuromyelitis optica
NRF	Nuclear respiratory factors
NSAID	Nonsteroidal anti-inflammatory drug
NuMTs	Nuclear mtDNA sequences
OAE	Otoacoustic emission
OCT	Optical coherence tomography
OGTT	Oral glucose tolerance test
OM	Outer membrane

OPMD	Oculopharyngeal muscular dystrophy
OSCP	Oligomycin sensitivity-conferring protein
OXPHOS	Oxidative phosphorylation
PCR	Polymerase chain reaction
PDHc	Pyruvate dehydrogenase complex
PEG	Percutaneous endoscopic gastrostomy
PEO	Progressive external ophthalmoplegia
PERG	Pattern electroretinography
PGD	Pre-implantation genetic diagnosis
PICU	Pediatric intensive care unit
PN	Parenteral nutrition
PNDE	Progressive neuronal degeneration of childhood with epilepsy
POLG	Polymerase gamma
PPAR	Peroxisomal proliferation activator receptors
PPM	Permanent pacemaker
PR	Pigmentary retinopathy
PRL	Prolactin
PS	Pearson syndrome
PTA	Pure-tone audiometry
PTH	Parathyroid hormone
RAPD	Relative afferent pupillary defect
RBP	Retinol-binding protein
RC	Respiratory chain
RGC	Retinal ganglion cells
RHADS	Rhythmic high-amplitude delta with superimposed (poly)spikes
RHODOS	Multicenter, double-blind, randomized, placebo-controlled trial
ROS	Reactive oxygen species
RPE	Retinal pigment epithelium
RRF	Ragged red fibers
SANDO	Sensory ataxia with neuropathy dysarthria and ophthalmoplegia
SANO	Sensory ataxic neuropathy and ophthalmoparesis
SCAE	Spinocerebellar ataxia with epilepsy
SDH	Succinate dehydrogenase
SLTP	Speech and language therapist
SMA	Superior mesenteric artery
SNHL	Sensorineural hearing loss
SNIP	Sniff nasal inspiratory pressures
SRNS	Steroid-resistant nephrotic syndrome
TCA	Tricarboxylic acid
TDF	Tenofovir
tRNA	Transfer RNA
TEOAE	Transient-evoked otoacoustic emission
TFAM	Transcription factor A
TFT	Thyroid function tests
TRMA	Thiamine-responsive megaloblastic anemia syndrome
TSH	Thyroid-stimulating hormone
TYMP	Thymidine phosphorylase

U&Es	Urea and electrolytes
VEP	Visual-evoked potential
VFSS	Videofluorographic swallowing study
VLCADD	Very-long-chain acyl-CoA dehydrogenase deficiency
VT	Ventricular tachycardia
WGS	Whole genome sequencing
WPW	Wolff-Parkinson-White syndrome

Chapter

1

Mitochondria in Health and Disease

Carlo Viscomi and Massimo Zeviani

Mitochondrial Structure and Function

Mitochondria are essential organelles found in nearly all eukaryotic cells [1]. The most prominent role for mitochondria is to supply the cell with spendable energy in the form of adenosine triphosphate (ATP) generated by oxidative phosphorylation (OXPHOS). Besides this fundamental role, mitochondria take part in a number of processes relevant for cellular physiology, including, among others, heat production, apoptosis, generation and detoxification of reactive oxygen species (ROS), intracellular Ca^{2+} regulation, steroid hormone and heme synthesis and lipid metabolism [2] (Figure 1.1).

Mitochondria are organized into four morphologically and functionally distinct compartments (Figure 1.2): (i) the outer membrane (OM), which is freely permeable to ions and small molecules, unlike bigger metabolites and proteins, whose traffic is mediated by specific transporters and channels; (ii) the intermembrane space (IMS), where several important processes take place, such as exchange of proteins, lipids or metal ions between the matrix

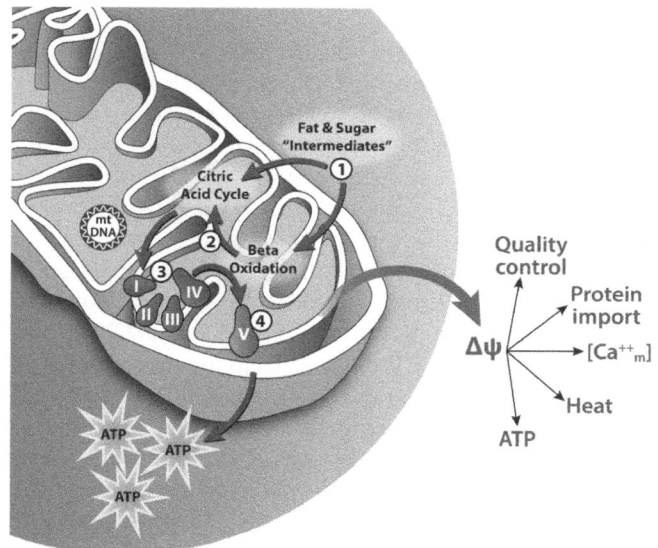

Figure 1.1 Main metabolic pathways within mitochondria. (A black and white version of this figure will appear in some formats. For the color version, please refer to the plate section.)

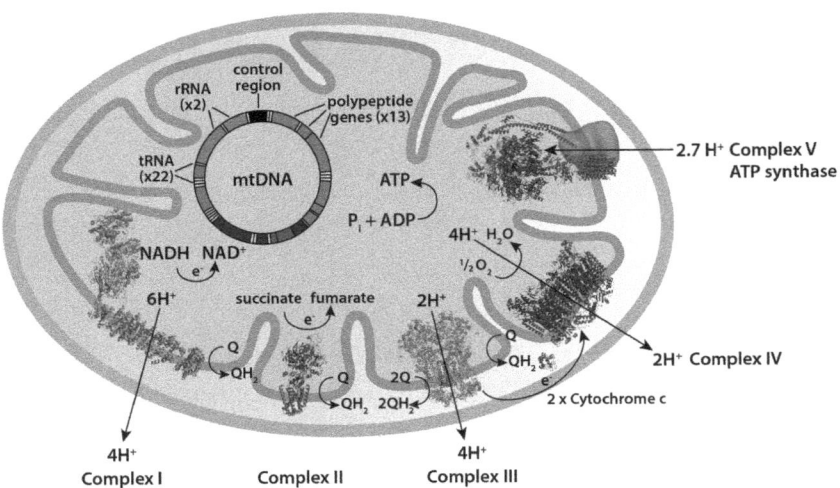

Figure 1.2 Overview of the OxPhos system (complexes I–V) and of the organization of mitochondrial DNA. (A black and white version of this figure will appear in some formats. For the color version, please refer to the plate section.)

and the cytosol, the initiation of apoptotic cascades and insertion of specific proteins, e.g., mitochondrial carriers, into the outer or inner membranes; (iii) the inner membrane (IM), which surrounds the matrix and mediates the transport of ions, metabolites and proteins through specialized transporters and folds into invaginations called cristae, where the respiratory complexes are localized; and (iv) the matrix, which contains the mtDNA and proteins involved in a huge number of biochemical pathways, including the TCA cycle and beta-oxidation of fatty acids (Figure 1.1).

Mitochondria harbor the complexes of the respiratory chain (RC), which convert the energy derived from nutrients into ATP (and heat) (Figure 1.2). This function is carried out by two coupled reactions, respiration and phosphorylation. Respiration is performed by four multiheteromeric complexes (CI–IV), which transfer the electrons extracted from the carbon substrates of nutrients along a redox potential to eventually reduce molecular oxygen into water. The energy liberated during these sequential redox reactions is exploited by proton pumps incorporated in RC complexes I, III and IV to translocate protons from the matrix to the IMS, thus forming an electrochemical membrane potential (ΔP) composed of a chemical (ΔpH) and an electrical ($\Delta\Psi$) gradient. ΔP is eventually exploited by the mitochondrial, oligomycin-sensitive ATP synthase (or complex V, CV) to convert ADP into ATP through a rotary reaction by which a complete rotation of the CV c-ring leads to the condensation of three ADP+Pi molecules into three molecules of ATP (Figure 1.2). The number of protons necessary for one rotation to complete varies according to the number of c subunits composing the ring rotor, which establishes the gear for the rotation to take place. In humans, the c-ring is composed of eight subunits, consenting the passage of as many protons through the rotor for each complete rotation. The whole process comprising both respiration and ATP biosynthesis is called oxidative phosphorylation. The respiratory complexes are composed of several subunits, require additional assembly factors and chaperons and are organized in supercomplexes, whose functional relevance is a matter of intense debate [3].

Complex I (CI), or NADH:ubiquinone oxidoreductase is the biggest among the RC complexes (molecular mass: 969 kDa) and transfers electrons from NADH to CoQ [4]. It forms an L-shaped structure with a peripheral arm protruding into the matrix and containing the NADH dehydrogenase and electron flux activities, and a membrane arm containing the proton pumps. The "heel" of the L contains a long channel, which accommodates CoQ. The protein backbone of CI is composed of 14 core subunits, seven of which are highly hydrophobic mtDNA-encoded proteins (all part of the membrane arm), and seven hydrophilic nuclear DNA (nDNA)-encoded subunits (part of the peripheral arm). In addition, human CI contains 30 supernumerary nDNA-encoded subunits, which play important, albeit still poorly understood roles for assembly, regulation, stability or protection against oxidative stress. Nonprotein components of CI include a flavin mononucleotide (FMN) moiety acting as the catalytic center for the NADH dehydrogenase activity of the complex, and eight Fe-S clusters which transfer electrons through the peripheral arm to CoQ.

Complex II (CII), or succinate:ubiquinone oxidoreductase, is assembled from four nDNA-encoded polypeptides (SDHA-D, molecular mass: 120 kDa), which transfer electrons from $FADH_2$ to coenzyme Q [5]. In addition to the FAD moiety bound to the largest (70 kDa) subunit, it contains three Fe-S centers carrying out the electron flux through the complex, and a poorly understood heme moiety embedded in the hydrophobic portion anchoring CII to the IM.

Complex III (CIII), or ubiquinol:cytochrome c oxidoreductase, is a homodimeric complex (molecular mass: 480 kDa) transferring electrons from CoQ to cytochrome c, and consists of 11 subunits for each monomer [6]. Cytochrome b is the only mtDNA-encoded subunit, and forms the catalytic core along with the Fe_2S_2 cluster-containing Rieske protein, and cytochrome c_1. Most of the other eight subunits are small proteins that surround the metalloprotein nucleus.

Complex IV (CIV), or cytochrome c oxidase (COX), is the terminal enzyme of the electron transfer chain, and catalyzes the electron transfer from reduced cytochrome c to molecular oxygen [7]. Mammalian CIV is a heteromeric complex composed of 14 subunits. The three largest, Cox1, Cox2 and Cox3, are highly hydrophobic transmembrane proteins encoded by mtDNA and form the catalytic core, while the 11 nDNA-encoded, smaller subunits are involved in the regulation of COX activity, its stability and dimerization of the catalytically active enzyme. Two copper centers, CuA and CuB, are contained in Cox2 and Cox1, respectively, whereas Cox1 also contains two heme a moieties, which are also part of the CIV catalytic core.

Complex V (CV), or ATP synthase, catalyzes the synthesis of ATP from ADP and inorganic phosphate (P_i) using the energy provided by the proton electrochemical gradient [8]. ATP synthase consists of the F_1 portion, a soluble portion situated in the mitochondrial matrix, composed of three copies of subunits α and β, and one copy of subunits γ, δ and ε, and the F_o portion, bound to the inner mitochondrial membrane consisting of eight c-ring subunits and one copy each of subunits a, b, d, F6 and the oligomycin sensitivity-conferring protein (OSCP). Subunits b, d, F6 and OSCP form the peripheral stalk that lies to one side of the complex. A number of additional subunits (e, f, g and A6 L), all spanning the membrane, are associated with F_o. Two of the F_o subunits, subunit a and A6 L, are encoded by the mtDNA ATP6 and ATP8 genes, respectively.

Mitochondrial Biogenesis and Quality Control

Mitochondria are under the double genetic control of mtDNA and nDNA, and a finely tuned genetic network has evolved in order to functionally connect the two genomes [9]. The pathways controlling mitochondrial biogenesis are centered on the activity of the PPARγ coactivators (PGC)1-α and -β, which interact and drive the activity of several OXPHOS-related transcription factors, including the Nuclear Respiratory Factors (NRF1 and 2), and the Peroxisomal Proliferator Activator receptors (PPAR, β, and γ) among others. NRFs and PPARs in turn increase the transcription of OXPHOS and fatty acid oxidation (FAO)–related genes. They also regulate the expression of mitochondrial transcription factor A (TFAM), which is an indispensable component of mitochondrial transcription and replication systems. PGC1α is the best-characterized member of this family, its activity being repressed by acetylation, operated by acetylase GCN5 and increased by deacetylation, mainly through the nuclear deacetylase SIRT1; and by phosphorylation, regulated by several kinases, including p38 MAPK, glycogen synthase kinase 3b (GSK3b) and AMP-dependent kinase (AMPK).

Mitochondria are highly dynamic organelles whose shape and network is regulated by the opposing processes of fusion and fission [10]. Mitodynamics transactions provide an important quality-control mechanism, since fusion contributes to mitochondrial maintenance, and fission allows the segregation and eventually disposal of dysfunctional mitochondria [11]. Dynamin-related GTPases on the OM (Mitofusins, MFN 1 and 2) and IM (OPA1) control the fission process, while the cytosolic soluble dynamin-related protein 1 (DRP1) regulates fission. DRP1 interacts with docking adaptors (FIS1, MFF and MiD49/51), forming spiral filaments that drive mitochondrial constriction and fragmentation upon translocation to the OM. This process is particularly important for the elimination of dysfunctional mitochondria that are then targeted for autophagic degradation, a process called mitophagy. Mitophagy can also be triggered by mitochondrial dysfunction through a decrease in membrane-potential-driven protein import. In normal conditions, the kinase PINK1 is imported into mitochondria and rapidly degraded; a decrease in import caused by ΔP drop causes PINK1 to accumulate on the outer membrane, where it recruits the E3 ligase Parkin, which in turn ubiquitinates a specific subset of OM proteins including Mitofusins, and promotes their proteasomal degradation. The Parkin-dependent degradation of factors involved in mitochondrial motility and fusion enhances the selectivity for the removal of defective mitochondria by mitophagy. Contrariwise, mitochondrial hyperfusion observed during nutrient starvation is supposed to protect mitochondria from mitophagy through steric hindrance [12].

Genetic Basis of Mitochondria (Including Inheritance)

The human mitochondrial genome is a double-stranded 16.6 kb circular DNA molecule (mtDNA) encoding 13 proteins, which are all part of four (CI, CIII, CIV, CV) of the five canonical multiheteromeric enzyme complexes constituting the OXPHOS system [13]. In addition, mtDNA contains genes encoding 22 tRNAs, and 12S and 16S rRNAs that are required for mitochondrial protein synthesis. There are no introns in the mitochondrial genome, and all 37 genes are adjacent to each other with few exceptions. The distribution of genes in the two mtDNA strands (called the heavy (H) and light (L) strands because of the different content in G-T versus C-A nucleotide residues) is asymmetrical. Both genes encoding the 12S and 16S ribosomal RNA and most of the tRNA- and protein-encoding genes are contained in the H strand, whereas the L strand contains only the ND6 gene, in addition to some tRNA-encoding genes. An untranslated region of approximately 1 Kb harbors the replication origin of the heavy strand and the promoters for transcription of both strands. Nuclear DNA (nDNA) genes encode all the other subunits which take part in the OXPHOS complexes, and also all the proteins required for their assembly, those carrying out the maintenance and expression of mtDNA, the biosynthesis of the respiratory cofactors and prosthetic groups etc. Thus, the mitochondrial proteome includes approximately 1,500 nDNA-encoded mitochondrial genes, in addition to the 37 mtDNA genes [14].

The intricate molecular machinery involved in the maintenance, transcription and translation of mtDNA includes several proteins encoded by nDNA [15]. These proteins are essential for mtDNA replication (POLγA and B, TWINKLE), for transcription/translation (TFAM, TFB1 M, mTERFs, LRPPRC) or for the balanced maintenance of the mitochondrial dNTPs pool (TP, TK, ANT1 and RRM2B).

Each human cell has hundreds to several thousand mitochondria and every mitochondrion can carry as many as 10 copies of mtDNA packed in nucleoprotein structures called nucleoids. The main protein component of nucleoids is TFAM, which acts as a DNA-packaging molecule by thoroughly binding the mtDNA molecule with a stoichiometry of 35–37 base-pairs per TFAM dimer. Cells and tissues with higher ATP demand typically have more mtDNA.

Usually, all mtDNA copies are identical, a condition known as homoplasmy. However, errors occurring during mtDNA replication or repair can lead to the formation of a mutant mtDNA molecule, which can clonally expand through unknown mechanisms, and eventually fixate in a metastable condition referred to as heteroplasmy, where mutant and wild-type genomes coexist in the same organelles/cells/tissues, in different proportions. Low levels of heteroplasmy have been shown to be present in normal cells, particularly in post-mitotic tissues such as skeletal muscle, or in stem cells of the colonic crypts, and increase over time in parallel with the aging of the individual. However, only when the mutation load of mtDNAs offsets a minimum critical threshold, usually ranging between 70 percent and 90 percent, does mitochondrial dysfunction become manifest in a particular tissue, leading to organ failure and development of a mitochondrial disease. Different tissues exhibit variation in their mutant threshold, with germ cells, for example, having minimal tolerance for the accumulation of mtDNA mutations.

Genetic Basis and Mechanisms of Mitochondrial Diseases

As a result of the dual genetic control of mitochondrial OXPHOS, genetic defects affecting mtDNA or OXPHOS-related nuclear DNA (nDNA) genes can compromise ATP synthesis, determine mitochondrial dysfunction and cause human disease [16]. Mitochondrial disorders are in fact defined as clinical entities associated with defects of mitochondrial OXPHOS, which are ultimately genetically determined. They can exhibit any kind of transmission, including maternal, autosomal dominant, autosomal recessive and X-linked modes of inheritance. While different gene mutations can give rise to a similar range of phenotypes, mutations in the same gene can often lead to a variety of different clinical entities. In addition, different levels of heteroplasmy of the same mtDNA mutation can result in a wide spectrum of phenotypes. Hundreds of pathogenic mtDNA mutations have been documented (MITO-MAP 2012) that can affect virtually every tissue in the body, leading to different phenotypes depending on their intrinsic severity, targeted gene and heteroplasmy levels. Tissue and organ functions critically depend on adequate ATP production, especially when energy demand is high, like in neurons and muscle fibers [17]. This explains why primary disorders of mitochondrial bioenergetics usually cause neurodegeneration and/or muscle weakness, leading to neuromuscular disease in children and adults. However, specific mitochondrial syndromes can involve any other organ, either individually or in combination with brain and muscle dysfunction. Many mitochondrial disorders also involve multiple organs causing cardiomyopathy and heart conduction defects, liver dysfunction, diabetes mellitus, sensorineural deafness, ophthalmoparesis (weak eye muscles), ptosis (drooping eyelids) and optic neuropathy causing blindness due to degeneration of the optic nerve. Mutations in mtDNA can affect specific proteins of the respiratory chain or the synthesis of mitochondrial proteins as a whole (when mutations or deletions involve tRNA or rRNA genes) and can be in turn divided into large-scale rearrangements (i.e. partial deletions or duplications) and inherited point mutations. Both groups have been associated with well-defined clinical syndromes. While large-scale rearrangements are usually sporadic, point mutations are typically maternally inherited. Large-scale rearrangements include several genes and are invariably heteroplasmic. In contrast, point mutations may be heteroplasmic or homoplasmic, the latter characterized by incomplete penetrance (e.g. Leber's Hereditary Optic Neuropathy).

An increasing number of mutations in nuclear genes encoding mitochondrial proteins have been described, leading to a range of different syndromes.

A genetic classification of mitochondrial disorders is presented in Table 1.1.

Table 1.1 Genetic classification of nDNA-related mitochondrial diseases

mtDNA mutations	Large-scale rearrangements of mtDNA
	Point mutations of mtDNA
nDNA mutations	Genes encoding structural subunits of the OXPHOS complexes
	Genes encoding factors affecting mtDNA maintenance, transcription and translation
	Genes encoding factors involved in the biosynthesis of lipids and cofactors
	Genes encoding proteins involved in mitochondrial protein import and dynamics
	Genes encoding assembly factors of the OXPHOS complexes
	Genes encoding enzymes involved in detoxification pathways
	Genes encoding factors involved in protein quality control
	Genes encoding factors involved in mitodynamics
	Genes encoding factors involved in apoptosis
	Genes encoding factors involved in ion transport
	Genes encoding factors involved in protein import

Mutations in mtDNA

Large-scale rearrangements of mtDNA: mtDNA-rearrangement syndromes are invariably heteroplasmic (homoplasmic large deletions being incompatible with life), and can result in a range of clinical manifestations and a wide spectrum of severity. The size of deletions can vary from a few hundred bases to several kilobases and several genes are usually involved. The syndromes associated with rearrangement of mtDNA range from maternally inherited type 2 diabetes and deafness due to an mtDNA-duplication mutation, to adult-onset chronic progressive external ophthalmoplegia (PEO), childhood or juvenile onset multisystem Kearns-Sayre syndrome (KSS) or a perinatal, life-threatening condition, Pearson's syndrome (PS). PEO is characterized by a progressive paralysis of the eye muscles, leading to impaired eye movement and bilateral drooping eyelids (ptosis). CPEO is typically caused by sporadic large-scale single deletions or multiple mtDNA deletions. KSS is characterized by early onset (childhood or young adulthood) of progressive external ophthalmoplegia, ptosis, mitochondrial myopathy with ragged red fibers, CNS involvement (progressive ataxia and cognitive decay) and potentially life-threatening abnormalities of the cardiac rhythm. PS is characterized by severe, usually fatal pancytopenia and insufficiency of exocrine pancreas. Interestingly, children surviving the pancytopenic phase of PS show a rapid decrease of deleted mtDNA species in bone marrow cells, their accumulation in skeletal muscle (and brain) and the evolution of the clinical features into early-onset KSS. The majority of single large-scale rearrangements of mtDNA is sporadic and therefore believed to be the result of the clonal amplification of a single mutational event, occurring in the maternal oocyte or early during the development of the embryo.

Point mutations of mtDNA: In contrast to large-scale rearrangements, mtDNA point mutations are usually maternally inherited. Mutations are considered pathogenic if they affect highly conserved nucleotide/amino acid, segregate with phenotype, show quantitative correlation between heteroplasmy and phenotype severity and are present in affected families from ethnically distinct human populations.

Mutations have been found in all mtDNA-encoded genes. The most common syndromes include: (i) mitochondrial encephalomyopathy, lactic acidosis, stroke-like episodes (MELAS), mainly due to mutations in the tRNA$^{\text{Leu(UUR)}}$ gene; (ii) myoclonus epilepsy and ragged red fibers (MERRF) due to mutations in the tRNA^{Lys} gene; (iii) Leber's hereditary optic neuropathy (LHON) due to mutations in the CI encoding genes ND1, ND4 and ND6; and (iv) neurogenic muscle weakness, ataxia, retinitis pigmentosa (NARP) due to mutations in the ATP6 gene. Different clinical presentations are also associated with mutations of CIII and CIV mtDNA-encoded subunits. In particular, *CYTB* mutations may lead to isolated myopathy, but also to a multisystem disorder characterized by encephalomyopathy, cardiomyopathy and septo-optic dysplasia; whereas mutations in *COX1*, *COX2* and *COX3* are associated with several manifestations, including MELAS, encephalomyopathy and motor neuron disease-like presentation.

Mutations in Nuclear Genes

Mutations in nDNA-encoded OXPHOS-related genes have also been linked to a variety of multisystem disorders [18]. Importantly, most of nuclear genes responsible for mitochondrial disease are not encoding structural subunits of the OXPHOS system, but rather ancillary factors involved in their assembly, activity and turnover, components of the mtDNA replication and expression machineries, enzymes controlling the intramitochondrial supply of deoxynucleotides, proteins involved in mitodynamics and quality control, in the biosynthesis of the lipid milieu or in the formation of prosthetic groups and cofactors (Table 1.1).

In spite of the substantial progress made in the past two decades in the molecular definition of numerous mitochondrial disorders and in the understanding of their pathophysiology, we still face major limitations for the development of new treatments and preventing these disorders in the near future [19].

First, in a large proportion of cases it is still not possible to reach a molecular genetic diagnosis. In fact, more than 50 percent of adult patients, and an even greater percentage of pediatric cases, remain undefined genetically [20], and the diagnosis is based on biochemical and/or morphological finding in muscle or, more rarely, in cultured fibroblasts. The lack of a genetic diagnosis prevents the patients from receiving reliable family counseling, prevents reliable prenatal diagnosis and necessitates referrals for biochemical methods that vary from lab to lab and are often incomplete. While the analysis of mtDNA is a well-standardized procedure in most Centers, the list of known disease genes associated with mitochondrial dysfunction is constantly increasing, and the identification of new genes has largely relied on the availability of large and/or multi-consanguineous families. New re-sequencing technologies offer the possibility of rapid and relatively low-cost characterization of whole genomes in individual patients or families, and is especially powerful when parent-child trios are sequenced in parallel. This enables the molecular dissection of mitochondrial disorders in humans at an unprecedented level.

Second, once a new mitochondrial disease gene has been identified, the process of validation of the mutant variants, and, even more importantly, the characterization of the function of the corresponding gene product, are essential steps for a complete understanding of the disease process. This often takes considerably longer than the initial genetic studies, and can limit progress, particularly when only one family has been identified with a possible pathogenic mutation in a new disease gene.

References

1. S. B. Vafai, V. K. Mootha. Mitochondrial disorders as windows into an ancient organelle, *Nature* 2012; **491**: 374–383.

2. D. C. Wallace. A mitochondrial paradigm of metabolic and degenerative diseases, aging, and cancer: a dawn for evolutionary medicine. *Annual Review of Genetics* 2005; **39**: 359–407.

3. M. L. Genova, G. Lenaz. Functional role of mitochondrial respiratory supercomplexes. *Biochimica et Biophysica Acta* 2014; **1837**: 427–443.

4. J. Hirst. Mitochondrial complex. *Annual Review of Biochemistry* 2013; **82**: 551–575.

5. H. J. Kim, O. Khalimonchuk, P. M. Smith, et al. Structure, function, and assembly of heme centers in mitochondrial respiratory complexes. *Biochimica et Biophysica Acta* 2012; **1823**: 1604–1616.

6. P. M. Smith, J. L. Fox, D. R. Winge. Biogenesis of the cytochrome bc(1) complex and role of assembly factors. *Biochimica et Biophysica Acta* 2012; **1817**: 276–286.

7. D. U. Mick, T. D. Fox, P. Rehling, Inventory control: cytochrome c oxidase assembly regulates mitochondrial translation. *Nature Reviews. Molecular Cell Biology* 2011; **12**: 14–20.

8. J. E. Walker. The ATP synthase: the understood, the uncertain and the unknown. *Biochemical Society Transactions* 2013; **41**: 1–16.

9. R. C. Scarpulla. Transcriptional paradigms in mammalian mitochondrial biogenesis and function. *Physiological Reviews* 2008; **88**: 611–638.

10. P. Mishra, D. C. Chan. Mitochondrial dynamics and inheritance during cell division, development and disease. *Nature Reviews. Molecular Cell Biology* 2014; **15**: 634–646.

11. R. J. Youle, D. P. Narendra. Mechanisms of mitophagy. *Nature Reviews. Molecular Cell Biology* 2011; **12**: 9–14.

12. L. C. Gomes, L. Scorrano. Mitochondrial morphology in mitophagy and macroautophagy. *Biochimica et Biophysica Acta* 2013; **1833**: 205–212.

13. E. A. Schon, S. DiMauro, M. Hirano. Human mitochondrial DNA: roles of inherited and somatic mutations. *Nature Reviews. Genetics* 2012; **13**: 878–890.

14. S. E. Calvo, V. K. Mootha. The mitochondrial proteome and human disease. *Annual Review of Genomics and Human Genetics* 2010; **11**: 25–44.

15. W. C. Copeland. Inherited mitochondrial diseases of DNA replication. *Annual Review of Medicine* 2008; **5**: 131–146.

16. M. Zeviani, S. Di Donato. Mitochondrial disorders. *Brain* 2004; **127**: 2153–2172.

17. W. J. Koopman, F. Distelmaier, J. A. Smeitink, et al. OXPHOS mutations and neurodegeneration. *The EMBO Journal* 2013; **32**: 9–29.

18. W. J. Koopman, P. H. Willems, J. A. Smeitink. Monogenic mitochondrial disorders. *New England Journal of Medicine* 2012; **366**: 1132–1141.

19. P. Chinnery, K. Majamaa, D. Turnbull, et al. Treatment for mitochondrial disorders. *The Cochrane Database of Systematic Reviews* 2006; CD004426.

20. S. E. Calvo, A. G. Compton, S. G. Hershman, et al. Molecular diagnosis of infantile mitochondrial disease with targeted next-generation sequencing. *Science Translational Medicine* 2012; **4**: 118ra110.

Chapter 2

Clinical Approach in Children

Shamima Rahman

Clinical History

Childhood-onset mitochondrial disease is diagnostically challenging since it is characterized by extreme clinical heterogeneity (Figure 2.1). Clinical suspicion of mitochondrial disease in an infant or a young child usually arises in one of three scenarios: 1) the child presents a constellation of symptoms and signs falling within the spectrum of a recognizable "classical" mitochondrial syndrome (see later in this chapter); 2) elevated lactate levels have been noted in blood, urine or cerebrospinal fluid (CSF); 3) a noncanonical multisystem presentation cannot be explained by any other disease process. A detailed history should be obtained, taking particular consideration of the following.

It is important to pay attention to the birth and neonatal history. Was there any unexplained anemia, jaundice or hypoglycemia? Did the patient feed well and gain weight appropriately in early life or was there gastroesophageal reflux, vomiting, diarrhea or constipation? Evidence of multisystem manifestations should be specifically sought; for

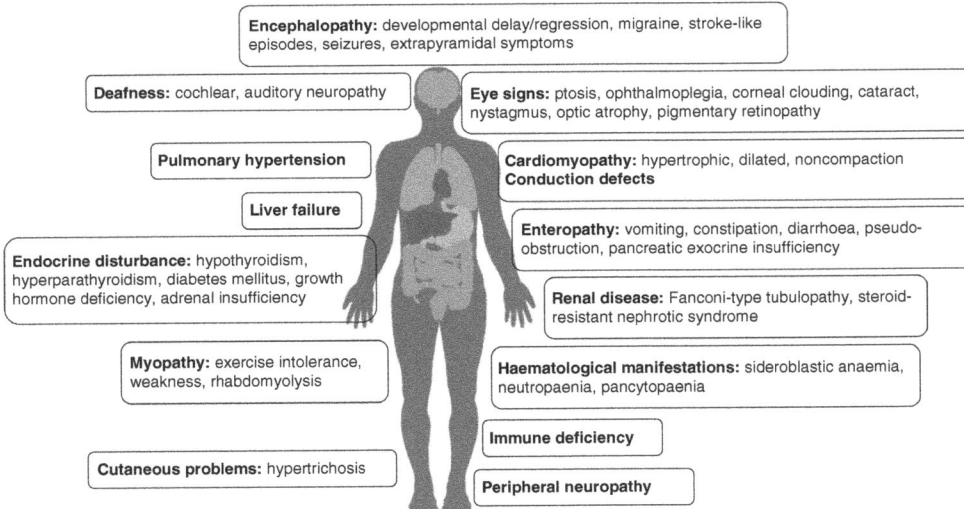

Figure 2.1 Pediatric mitochondrial disease is clinically extremely heterogeneous, and any combination of organ systems may be affected with onset at any age. This figure illustrates some of the more commonly recognized clinical manifestations of mitochondrial disease in childhood.

example, is there any history of diabetes mellitus, hypothyroidism, adrenal insufficiency, other endocrine disturbance, hearing loss, visual impairment, ptosis, squint, cataract, migrainous headache, stroke-like episodes, muscle cramps or red or dark urine? What is the child's exercise tolerance? Can they keep up with their peers in school sports? What kind of physical activities do they prefer? Are symptoms worse when the child is fatigued?

The tempo of developmental progress is also important. Many children with mitochondrial disease have characteristic stepwise regression of neurodevelopmental skills following intercurrent illnesses or other metabolic stresses, typically with incomplete recovery so that the affected child never regains their former skill level. What medications have been tried and was there any response to these? Were there any adverse reactions to medications or anesthetic agents?

Family history is important, asking specifically about similar or discordant symptoms compatible with mitochondrial disease in other family members, any consanguinity or evidence of matrilineal inheritance. A matrilineal history of symptoms may suggest a mitochondrial DNA (mtDNA) mutation, whereas consanguinity raises suspicion of autosomal recessive inheritance. However, individuals from consanguineous families harboring mtDNA mutations have been well documented in the literature. Furthermore, mitochondrial disease can be inherited in several different ways (maternal, autosomal recessive, autosomal dominant, X-linked) or be sporadic, and the majority of cases do not have any relevant family history.

Clinical Examination

A thorough physical examination should include auxology and full neurological examination, searching for neurological signs and evidence of multisystem involvement. Neurological features may include ptosis, ophthalmoplegia, hypotonia, spasticity, dystonia, muscle weakness (particularly proximal) or cerebellar signs. However, it is quite common for muscle power to be normal on static testing in mitochondrial disease. Fundoscopy may reveal pallor of the optic disc or pigmentary retinopathy. Tachypnea may reflect lactic acidosis, and a cardiac murmur may be audible in cases with cardiomyopathy. Hepatomegaly is an unusual feature in mitochondrial disease, but may occur in some of the mitochondrial DNA depletion syndromes (MDDS) [1].

Clinical Investigations

Clinical investigations in suspected mitochondrial disease aim to delineate the extent of multisystem involvement. Electromyography and nerve conduction studies may reveal myopathic or neuropathic changes. Expert ophthalmological review is important to identify pigmentary retinopathy and optic atrophy, while audiological investigations may reveal sensorineural hearing loss (SNHL) due to cochlear dysfunction or auditory neuropathy spectrum disorder. Cardiac investigations should include electrocardiogram to screen for conduction defects, as well as echocardiography, which may reveal hypertrophic cardiomyopathy (HCM), noncompaction of the left ventricle or dilated cardiomyopathy (DCM, e.g. in Barth syndrome or in the late stages of HCM). Blood investigations to screen for multisystem involvement should include full blood count, liver function, glucose, thyroid function, cortisol and parathyroid hormone levels. Measurement of urinary tubular proteins such as N-acetylglucosaminidase and retinol binding protein are helpful in identifying renal tubular dysfunction, together with calcium/creatinine and magnesium/creatinine ratios.

Brain magnetic resonance imaging may be helpful in the diagnostic process, by identifying characteristic lesions, as in Leigh and MELAS syndromes [2,3]. The former is typified by bilateral symmetrical lesions in the basal ganglia, variably extending through the midbrain to the brainstem. These lesions appear hyperintense on T1 weighted and FLAIR imaging, but hypointense in T2 sequences and as hypodensities on computed tomography [2]. Patients with MELAS typically have parieto-occipital lesions that do not correspond to vascular territories [3]. Cystic leukoencephalopathies are increasingly recognized in various forms of mitochondrial disease, including some subtypes of complex I deficiency and other forms of Leigh syndrome. A few specific gene defects are associated with virtually pathognomonic MRI brain changes, e.g. leukoencephalopathy with brain stem and spinal cord involvement and lactate elevation (LBSL) with *DARS2* mutations, leukoencephalopathy with thalamus and brainstem involvement and high lactate (LBTL) with *EARS2* mutations and pontocerebellar hypoplasia type 6 with *RARS2* mutations. Magnetic resonance spectroscopy may be helpful in revealing a lactate peak, although this is not specifically diagnostic. Identification of a succinate peak suggests succinate dehydrogenase deficiency.

Metabolic investigations should include blood lactate and pyruvate levels, ideally in free-flowing samples deproteinized at the bedside. CSF lactate may be elevated even when blood levels are normal. Plasma amino acids may reveal elevated alanine (reflecting lactic acidosis), high proline or low citrulline in some mitochondrial disorders. Urinary organic acids typically reveal lactate and Krebs cycle metabolites and are thus usually not specifically diagnostic, but occasionally contain useful metabolic clues such as the presence of 3-methylglutaconic acid (e.g. in MEGDEL, Barth and Sengers syndromes – see later in this chapter), methylmalonic acid in succinylCoA ligase deficiency or increased levels of ethylmalonic acid, methylsuccinic acid and isobutyryl-, isovaleryl-, 2-methylbutyryl- and hexanoylglycine in ethylmalonic encephalopathy. Metabolic investigations may also be helpful in identifying other metabolic causes of complex multisystem disease, e.g. very long chain fatty acids may be abnormal in peroxisomal disorders, while transferrin electrophoresis may reveal atypical transferrin glycoforms in congenital disorders of glycosylation.

Specific biochemical and genetic investigations for mitochondrial disease are complex and will be discussed in other chapters, but Figure 2.2 outlines a suggested diagnostic approach for the investigation of suspected mitochondrial disease in childhood.

Classical Clinical Syndromes (Limited to Those Presenting Predominantly in Childhood)

The following section discusses the more common clinical presentations of mitochondrial disease in childhood, arranged broadly according to age at presentation, starting with disorders with onset in the neonatal period and early infancy, followed by those presenting in later childhood and adolescence. The coenzyme Q_{10} (CoQ_{10}) biosynthesis defects are described at the end, since these are heterogeneous disorders with a broad range of onset, ranging from the neonatal period (multisystem disorders) to adolescence (ataxic and myopathic presentations) [4].

Congenital Lactic Acidosis

The earliest clinical presentation of mitochondrial disease is with congenital lactic acidosis, presenting in the neonatal period with tachypnea and nonspecific illness that may mimic

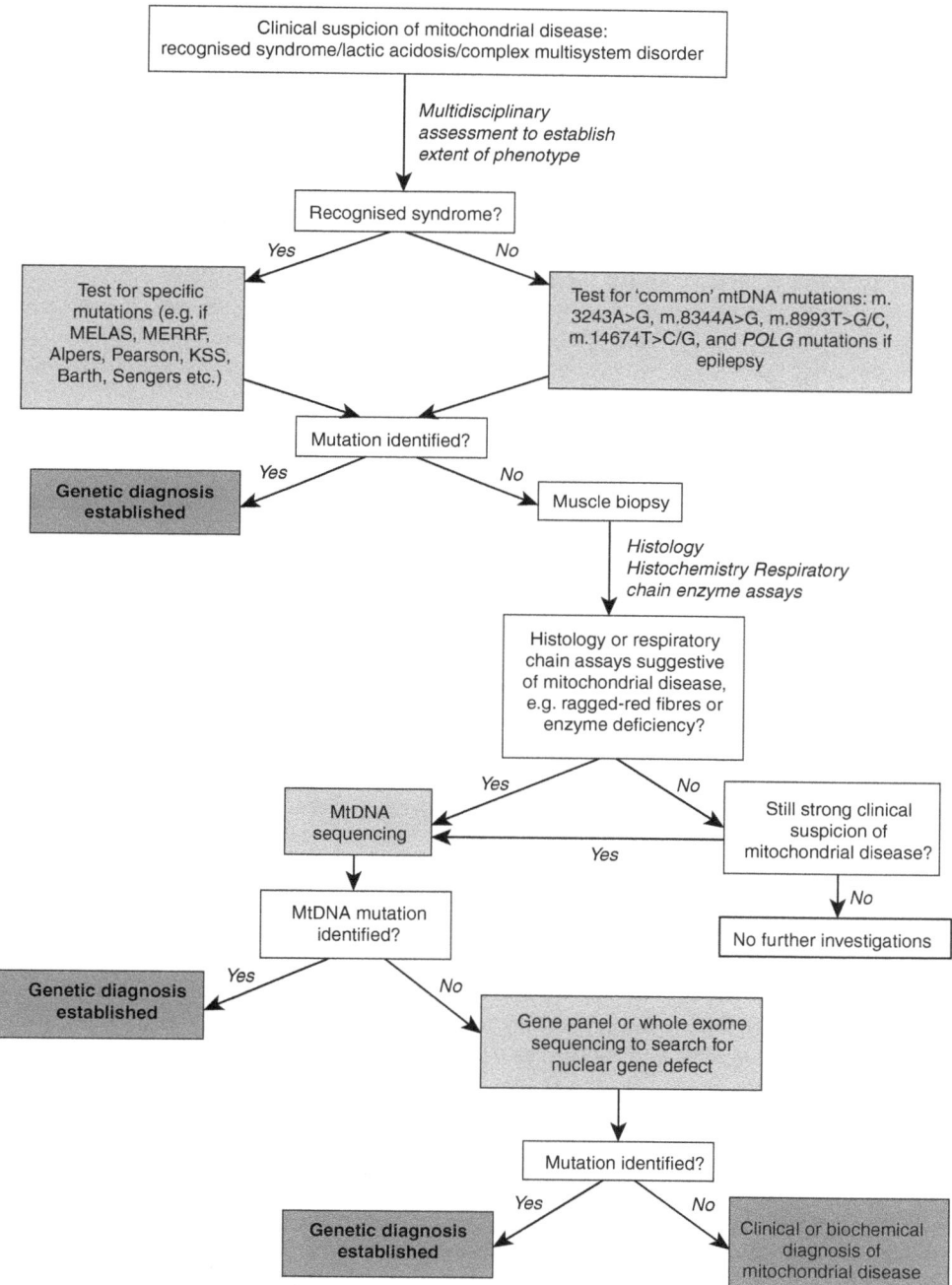

Figure 2.2 Flowchart illustrating a suggested diagnostic approach to the child with suspected mitochondrial disease

neonatal septicemia. There may be multisystem involvement, for example HCM, or features of Leigh syndrome (see later in this chapter). Causes of congenital lactic acidosis include pyruvate dehydrogenase deficiency, pyruvate carboxylase deficiency and isolated or multiple respiratory chain deficiencies. The treatable disorder biotinidase deficiency is classically associated with alopecia, dermatitis, hearing loss, optic atrophy and a characteristic organic aciduria. However, these are all late features, and biotinidase activity should be assayed in all infants with lactic acidosis and treatment with biotin given until the result is obtained. Lactic acidosis may also be secondary to Krebs cycle defects, organic acidemias (such as methylmalonic and propionic acidemias) and long chain fatty acid oxidation disorders. Acquired causes of lactic acidosis, such as sepsis, hypoxia, hypovolemia and other systemic illnesses, should be excluded (and appropriate treatment given), as should artefactual elevation of lactate arising from struggling or squeezing during blood sampling. A low lactate/pyruvate ratio is suggestive of pyruvate dehydrogenase deficiency, which is also often associated with cerebral malformations [5].

Leigh Syndrome

The most frequent childhood presentation of mitochondrial disease is Leigh syndrome, also known as subacute necrotizing encephalomyelopathy. Children with Leigh syndrome are well at birth and typically present in infancy with stepwise neurodevelopmental regression after an intercurrent viral illness (which may appear trivial) or other metabolic stress [2]. Later onset is recognized, including in adult life in occasional cases. Neurological features include hypo- or hypertonia (including spasticity and dystonia), ataxia, tremor, optic atrophy and ophthalmoplegia. Other systems may be affected, for example the gastrointestinal tract (vomiting, constipation, diarrhea, faltering growth) and the cardiovascular system (HCM). The clinical course is of periods of stability followed by further unpredictable episodes of deterioration, for which triggers may or may not be apparent. The late stages of the disease are characterized by progressive brainstem involvement, leading to loss of swallow, intractable vomiting in some cases and, eventually, death from central respiratory failure.

Originally Leigh syndrome was a neuropathological diagnosis, but nowadays diagnosis rests on the observation of characteristic MRI lesions (bilateral symmetrical lesions variably involving the basal ganglia and/or brainstem, which appear hyperintense on T2 weighted images), together with elevated lactate levels in blood and/or cerebrospinal fluid, in the clinical context described in the preceding paragraph. Leigh syndrome is biochemically and genetically heterogeneous, and can be caused by mutations in mtDNA or in more than 89 nuclear genes encoding mitochondrial proteins [6].

MEGDEL Syndrome

MEGDEL is a form of Leigh syndrome characterized by 3-methylglutaconic aciduria (MEG), deafness (D), encephalopathy (E) and Leigh-like disease (L) [7]. Clinical presentation typically occurs in the neonatal period or early infancy with hypoglycemia, lactic acidosis and hyperammonemia, often with hepatic dysfunction. The presentation may be mistaken for neonatal sepsis, and identification of increased urinary excretion of 3-methylglutaconic and 3-methylglutaric acid is key to making a prompt diagnosis. Subsequent problems include poor feeding and faltering growth in infancy, followed by the development of progressive SNHL, dystonia and spasticity. Profound developmental delay affects all domains, and affected children may have severe behavioral disturbance. MRI brain reveals bilateral basal ganglia lesions, together with cerebral and cerebellar atrophy. It has been suggested that MRI features in the early stages of the disease may be pathognomonic, with an "eye"-shaped region of signal sparing in the dorsal putamina. Causative mutations in the *SERAC1* gene affect phosphatidylglycerol remodeling in the mitochondrial membranes. Unlike many forms of Leigh syndrome, survival into the second decade may occur, although death may follow multiorgan failure in the neonatal period or liver failure in infancy.

Mitochondrial DNA Depletion Syndromes

The mtDNA depletion syndromes (MDDS) are a clinically and genetically heterogeneous group of autosomal recessive disorders of mtDNA maintenance characterized by progressive mtDNA depletion in affected tissues [1]. The causative genes encode components of the mtDNA replication apparatus or of mitochondrial nucleoside salvage. Clinical presentations of MDDS can be broadly classified into four main groups: hepatocerebral, myopathic, encephalomyopathic and neurogastrointestinal. Onset is usually in infancy and most, but not all, of these disorders lead to death in infancy or early childhood. Hepatocerebral MDDS typically presents in infancy with liver dysfunction associated with persistent vomiting, hypoglycemia, lactic acidosis and developmental delay and/or regression. Mutations in five genes may cause hepatocerebral MDDS: *DGUOK, POLG, C10orf2, MPV17* and *SUCLG1*. Recessive *POLG* and *C10orf2* mutations may present with hepatocerebral disease in infancy or with intractable epilepsy including Alpers-Huttenlocher syndrome (see later in this chapter). Recessive *MPV17* mutations are also associated with Navajo hepatopathy, consisting of liver disease, severe demyelinating sensorimotor neuropathy and leukoencephalopathy. Myopathic MDDS is caused by *TK2* mutations leading to thymidine kinase deficiency, and presents in infancy with hypotonia and muscle weakness. The bulbar musculature is also affected leading to feeding difficulties. Progressive weakness leads to death from respiratory failure in early childhood, although occasionally patients may survive to their teenage years. Creatine kinase is typically markedly elevated, distinguishing TK2 deficiency from other mitochondrial disorders presenting in infancy and early childhood. Rarely patients with *TK2* mutations may present with hepatocerebral disease. Encephalomyopathic MDDS presents with global developmental delay, hypotonia and muscle weakness in infancy, and may be associated with SNHL, dystonia, Leigh-like MRI lesions and methylmalonic aciduria (mutations in *SUCLA2* encoding the ADP-forming beta subunit of succinylCoA ligase); fatal infantile lactic acidosis with methylmalonic aciduria (mutations in *SUCLG1* encoding the alpha subunit of succinylCoA ligase); with prominent renal involvement (mutations in *RRM2B* encoding the p53R2 subunit of ribonucleotide reductase); or with seizures and a movement disorder (mutations in *ABAT* encoding the GABA transaminase) [8]. Mutations in the most recent gene linked to MDDS, *FBXL4*, are associated with facial dysmorphism, skeletal abnormalities, poor growth, gastrointestinal dysmotility, renal tubular acidosis and seizures. The neurogastrointestinal presentation of MDDS is known as mitochondrial neurogastrointestinal encephalomyopathy (MNGIE) and is usually caused by *TYMP* mutations leading to thymidine phosphorylase deficiency (see later in this chapter).

Alpers-Huttenlocher Syndrome

Alpers-Huttenlocher syndrome (also known as progressive neuronal degeneration of childhood with epilepsy, PNDE) is an autosomal recessive multisystem disorder usually caused by compound heterozygous *POLG* mutations leading to severe mtDNA depletion [9]. Recessive mutations in *C10orf2* may result in a similar phenotype. Clinical features consist of a classical triad of intractable seizures (typically resistant to multiple antiepileptic drugs), developmental regression and (terminally) liver failure. The seizures are often focal initially, with later generalization. Once seizures appear, the clinical course is frequently rapidly progressive, leading to death from intractable seizures or liver failure within weeks to months. However, in some cases, disease progression may be slower, and occasionally there may be no further progression for several years. Liver failure may be triggered by sodium valproate therapy, but may also occur without exposure to valproate. EEG in the early stages of disease may be pathognomonic, showing unilateral occipital rhythmic high-amplitude delta with superimposed (poly)spikes (RHADS), although at later disease stages may be nonspecifically abnormal [10]. MRI brain may reveal posterior lesions, but lesions may also involve the deep cerebellar nuclei, thalamus and basal ganglia. In some patients, MRI may be normal or show only nonspecific changes.

Mitochondrial Neurogastrointestinal Encephalopathy Disease (MNGIE)

Symptom onset in MNGIE is often in childhood, although diagnosis is delayed until adult life in many cases. The first symptoms are usually related to gastrointestinal dysmotility (pseudo-obstruction, dysphagia, early satiety, nausea and vomiting after eating, bloating, constipation, diarrhea and severe cachexia), followed by later onset of ptosis and/or ophthalmoplegia, demyelinating peripheral neuropathy, SNHL and myopathy. MRI brain usually reveals leukoencephalopathy, which is typically asymptomatic [11]. The causative *TYMP* mutations lead to an inability to salvage nucleosides for mtDNA replication and hence to progressive mtDNA depletion as well as accumulation of mtDNA deletions and point mutations in tissues including nerves, muscle and brain. Some patients have been successfully treated with allogeneic stem cell transplantation, which restores thymidine phosphorylase activity. However, early treatment seems essential for a good outcome, since transplant-related morbidity and mortality appear to be directly related to the patient's disease severity at transplantation.

"Benign Reversible" Mitochondrial Myopathy

Infants with the so-called syndrome of "benign reversible" mitochondrial myopathy are usually normal at birth, but develop severe lactic acidosis from a few weeks of age associated with rapidly progressive myopathy characterized by hypotonia and severe muscle weakness. Affected infants generally need support with feeding (nasogastric or gastrostomy) and some need ventilatory support. Gradual recovery starts from approximately six months of age, and the need for artificial ventilation has usually resolved by 12–18 months. The disorder has been linked to two homoplasmic mtDNA point mutations, m.14674T>C and m.14674T>G in *MT-TE* encoding the tRNA for glutamic acid, but the molecular mechanisms underlying disease onset and spontaneous remission remain obscure [12].

Barth Syndrome

Barth syndrome, also known as 3-methylglutaconic aciduria type II, is an X-linked disorder characterized by DCM with endocardial fibroelastosis, skeletal myopathy (mainly proximal, leading to hypotonia, muscle weakness and fatigue), neutropenia (which may be cyclical) and growth delay [13]. HCM may occur, and left ventricular noncompaction. Causative mutations in the *TAZ* gene (also called *G4.5*) lead to decreased production of an enzyme tafazzin required to produce cardiolipin. Cardiolipin is a major lipid component of the mitochondrial inner membrane, and pathogenesis of Barth syndrome is thought to result from impaired energy metabolism because of instability of OXPHOS complexes in the abnormal lipid milieu of the mitochondrial membrane.

Sengers Syndrome

Sengers syndrome is a rare autosomal recessive disorder characterized by congenital cataract, HCM, skeletal myopathy, exercise intolerance and lactic acidosis. Mutations in the acylglycerol kinase (*AGK*) gene have been shown to be the underlying cause in many patients [14]. *AGK* encodes a step in the pathway of complex lipid biosynthesis in the mitochondrial membrane and, as for Barth syndrome, membrane instability is thought to lead to multiple OXPHOS defects. The same mechanism may be responsible for the severe mtDNA depletion reported in some patients. Not all patients with cataract and HCM have *AGK* mutations, suggesting that Sengers syndrome may be genetically heterogeneous. Outcome in Sengers syndrome is variable. Progressive HCM can lead to death from cardiac failure in infancy, but patients with long survival have been reported.

Pearson Syndrome

Children with the Pearson marrow pancreas syndrome usually present at a few weeks of age, after a period of relative "normality," with a severe transfusion-dependent anemia, typically associated with lactic acidosis [15]. Other hematopoietic lineages may be involved, so there may be neutropenia and/or thrombocytopenia in addition to the anemia. Analysis of bone marrow aspirates reveals ringed sideroblasts and vacuolation of myeloid precursors. The underlying cause is a heteroplasmic large-scale mtDNA deletion. A large number of different mtDNA deletions have been reported, but there is a "common" deletion of ~4.9kb. Affected children typically remain transfusion dependent for the first two years of life, but the transfusion requirement eventually tails off and completely resolves, mirroring clearance of the responsible mtDNA deletion from rapidly dividing blood cells. However, the clinical course is complicated by progressive multi-system disease (associated with accumulation of mtDNA deletions in nondividing tissues), frequently involving the kidneys (renal tubulopathy associated with severe electrolyte losses and sometimes progressing over time to end-stage renal failure), heart (cardiomyopathy and/or conduction defects including complete heart block), pancreas (exocrine and endocrine insufficiency), other endocrine glands (e.g. hypothyroidism, hypoparathyroidism, adrenal insufficiency) and the brain (see Kearns-Sayre syndrome, KSS, below). There is a high mortality in the first five years of life, typically resulting from liver failure or overwhelming acidosis. Those who survive inevitably develop the neurological features of KSS (see below).

Kearns-Sayre Syndrome

KSS is also caused by large-scale rearrangements of the mtDNA, often the common ~4.9kb deletion that may also cause Pearson syndrome (see above). KSS is defined clinically by the triad of progressive external ophthalmoplegia, pigmentary retinopathy and onset before 20 years, with at least one of the following additional features: heart block, cerebellar ataxia or raised cerebrospinal fluid (CSF) protein (>1g/l). Other clinical features include cardiac conduction defects (which may also occur in Pearson syndrome), short stature, cognitive deficits/mental retardation, dysphagia and diabetes mellitus. The causative mtDNA deletions may be detected in DNA extracted from blood or urine of KSS patients, but sometimes a muscle biopsy is needed to identify the pathogenic mutation, particularly in older children and adolescents. Deletions of mtDNA are usually sporadic but rarely may be maternally inherited. Furthermore, occasional cases of KSS with multiple mtDNA deletions and autosomal recessive inheritance have been reported, for example in association with mutations in *RRM2B*, a gene required for mtDNA maintenance [16].

MELAS Syndrome

The syndrome of mitochondrial encephalopathy with lactic acidosis and stroke-like episodes (MELAS) is a maternally inherited multisystem disorder [3]. MELAS usually manifests in childhood after a period of normal early development. The core clinical feature is stroke-like episodes, which often occur in late childhood or adolescence, although presentation may be later in adult life. The stroke-like episodes are typically heralded by migraine-like headache associated with vomiting and seizures. Hemianopia or cortical blindness may occur, and other clinical features include myopathy, cognitive decline, myoclonus, ataxia, episodic coma, optic atrophy, short stature, SNHL, HCM (which may lead to sudden death), nephropathy, gastro-intestinal dysmotility and diabetes mellitus. Approximately 80 percent of patients have a common mtDNA point mutation, m.3243A>G, in the *MT-TL1* gene encoding a transfer RNA (tRNA) for leucine. However, only a minority of patients with this mutation present with MELAS, and there can be great clinical variability, even between different members of the same family, possibly reflecting different mutation loads and tissue distribution of the mutation. The most frequent clinical presentation associated with the m.3243A>G mutation is with the syndrome of maternally inherited deafness and diabetes (MIDD), which usually manifests in adult life. The remaining 20 percent of patients with MELAS have other mtDNA mutations (www.mitomap.org). The diagnosis can be established in most cases by screening blood or urine DNA for the m.3243A>G mutation. However, sometimes a muscle biopsy may be needed to confirm the diagnosis, particularly in cases caused by other mtDNA mutations, where it may be necessary to sequence the entire mitochondrial genome in a muscle sample in order to identify the causative mutation.

MERRF Syndrome

Myoclonic epilepsy with ragged red fibers (MERRF) is a maternally inherited multisystem disorder characterized by myoclonus, generalized seizures, ataxia and ragged red fibers (RRF) in muscle [17]. Affected individuals may also develop SNHL, optic atrophy, pigmentary retinopathy, nystagmus, ophthalmoparesis, dysarthria, exercise intolerance, cardiomyopathy, lactic acidosis and multiple symmetrical lipomas. Symptom onset is usually in childhood or adolescence, but clinical presentation varies widely among affected individuals, even within a single pedigree. Approximately 80 percent of cases have a "common" mtDNA point mutation, m.8344A>G, in the *MT-TK* gene encoding the tRNA for lysine. MERRF has also been associated with other point mutations in mtDNA (www.mitomap.org), and some patients with *POLG* mutations may have similar clinical features. The diagnosis may be confirmed by screening blood or urine DNA for the m.8344A>G mutation. In some cases muscle biopsy may be needed.

Coenzyme Q_{10} Deficiency Syndromes

Disorders of coenzyme Q_{10} biosynthesis are clinically heterogeneous, and five major phenotypes are recognized: (1) encephalomyopathy, (2) severe infantile multisystemic disease, (3) cerebellar ataxia, (4) isolated myopathy and (5) steroid-resistant nephrotic syndrome (SRNS) [4]. Prompt diagnosis of disorders of coenzyme Q_{10} biosynthesis is important since early treatment with exogenous CoQ_{10} supplementation may result in a good outcome. Particular diagnostic clues suggestive of CoQ_{10} deficiency include ataxia and SRNS, which can occur in isolation or in association with SNHL, seizures and/or learning difficulties. The multisystem infantile-onset presentation of CoQ_{10} deficiency includes SNHL, optic atrophy, ataxia, dystonia, weakness and stroke-like episodes in addition to rapidly progressive nephropathy. However, because primary CoQ_{10} deficiency is potentially treatable, this should be in the differential diagnosis of all mitochondrial disease presenting in childhood, whatever the clinical features. Biochemical suspicion of CoQ_{10} deficiency may arise from the observation of deficiencies of complexes I+III and II+III in muscle biopsy, with normal activities of the individual complexes I, II and III when assayed separately, because the combined I+III and II+III assays are dependent on endogenous CoQ_{10} levels, whereas the individual assays are not. However, other patterns of respiratory chain enzyme deficiency may occur in primary CoQ_{10} deficiency, therefore direct measurement of CoQ_{10} in muscle biopsy is the preferred screening test. Treatment of primary CoQ_{10} deficiency is with high-dose CoQ_{10} supplementation in doses of at least 30mg/kg/day in childhood. Response to treatment is variable, ranging from complete prevention of symptoms if started early, to persistent ataxia or progressive renal impairment in other cases.

Isolated Organ Involvement

Although clinical suspicion of mitochondrial disease is greatest in the context of classical syndromic presentations or complex multisystem disorders, not all patients have classical syndromes or multisystem disease. It is therefore important to be vigilant for the possibility of mitochondrial disease in patients with single organ disease, including epileptic encephalopathy, myoclonic epilepsy and epilepsia partialis continua; HCM; acute liver failure; SRNS; sideroblastic anemia and pancytopenia. As more and more gene defects responsible for mitochondrial disease are identified, the range of phenotypes associated with these disorders is expanding exponentially (Figure 2.1).

References

1. Rahman S, Poulton J. Diagnosis of mitochondrial DNA depletion syndromes. Arch Dis Child. 2009 Jan;94(1):3–5. doi: 10.1136/adc.2008.147983. PubMed PMID: 19103785.

2. Rahman S, Blok RB, Dahl HH, Danks DM, Kirby DM, Chow CW, Christodoulou J, Thorburn DR. Leigh syndrome: Clinical features and biochemical and DNA abnormalities. Ann Neurol. 1996 Mar;39(3):343–351. PubMed PMID: 8602753.

3. Kaufmann P, Engelstad K, Wei Y, Kulikova R, Oskoui M, Sproule DM, Battista V, Koenigsberger DY, Pascual JM, Shanske S, Sano M, Mao X, Hirano M, Shungu DC, Dimauro S, De Vivo DC. Natural history of MELAS associated with mitochondrial DNA m.3243A>G genotype. Neurology. 2011 Nov 29;77 (22):1965–1971. doi: 10.1212/ WNL.0b013e31823a0c7f. Epub 2011 Nov 16. PubMed PMID: 22094475; PubMed Central PMCID: PMC3235358.

4. Emmanuele V, López LC, Berardo A, Naini A, Tadesse S, Wen B, D'Agostino E, Solomon M, DiMauro S, Quinzii C, Hirano M. Heterogeneity of coenzyme Q_{10} deficiency: Patient study and literature review. Arch Neurol. 2012 Aug;69 (8):978–983. doi: 10.1001/ archneurol.2012.206. Review. Erratum in: Arch Neurol. 2012 Jul;69(7):886. López, Luis [corrected to López, Luis C]. PubMed PMID: 22490322; PubMed Central PMCID: PMC3639472.

5. Horvath R, Kemp JP, Tuppen HA, Hudson G, Oldfors A, Marie SK, Moslemi AR, Servidei S, Holme E, Shanske S, Kollberg G, Jayakar P, Pyle A, Marks HM, Holinski-Feder E, Scavina M, Walter MC, Coku J, Günther-Scholz A, Smith PM, McFarland R, Chrzanowska-Lightowlers ZM, Lightowlers RN, Hirano M, Lochmüller H, Taylor RW, Chinnery PF, Tulinius M, DiMauro S. Molecular basis of infantile reversible cytochrome c oxidase deficiency myopathy. Brain. 2009 Nov;132 \(Pt 11):3165–3174. doi: 10.1093/brain/ awp221. Epub 2009 Aug 31. PubMed PMID: 19720722; PubMed Central PMCID: PMC2768660.

6. Patel KP, O'Brien TW, Subramony SH, Shuster J, Stacpoole PW. The spectrum of pyruvate dehydrogenase complex deficiency: Clinical, biochemical and genetic features in 371 patients. Mol Genet Metab. 2012 Jan;105(1):34–43. doi: 10.1016/j. ymgme.2011.09.032. Epub 2011 Oct 7. Review. Erratum in: Mol Genet Metab. 2012 Jul;106(3):384. Corrected and republished in: Mol Genet Metab. 2012 Jul;106(3): 385–94. PubMed PMID: 22079328; PubMed Central PMCID: PMC3754811.

7. Rahman J, Noronha A, Thiele I, Rahman S. Leigh map: A novel computational diagnostic resource for mitochondrial disease. Ann Neurol. 2017 Jan;81(1):9–16. doi: 10.1002/ana.24835. PubMed PMID: 27977873; PubMed Central PMCID: PMC5347854.

8. Wortmann SB, Vaz FM, Gardeitchik T, Vissers LE, Renkema GH, Schuurs-Hoeijmakers JH, Kulik W, Lammens M, Christin C, Kluijtmans LA, Rodenburg RJ, Nijtmans LG, Grünewald A, Klein C, Gerhold JM, Kozicz T, Van Hasselt PM, Harakalova M, Kloosterman W, Barić I, Pronicka E, Ucar SK, Naess K, Singhal KK, Krumina Z, Gilissen C, Van Bokhoven H, Veltman JA, Smeitink JA, Lefeber DJ, Spelbrink JN, Wevers RA, Morava E, de Brouwer AP. Mutations in the phospholipid remodeling gene SERAC1 impair mitochondrial function and intracellular cholesterol trafficking and cause dystonia and deafness. Nat Genet. 2012 Jun 10;44(7):797–802. doi: 10.1038/ ng.2325. PubMed PMID: 22683713.

9. Viscomi C, Zeviani M. MtDNA-maintenance defects: Syndromes and genes. J Inherit Metab Dis. 2017 Jul;40 (4):587–599. doi: 10.1007/s10545-017-0027-5. Epub 2017 Mar 21. PubMed PMID: 28324239; PubMed Central PMCID: PMC5500664.

10. Hikmat O, Tzoulis C, Chong WK, Chentouf L, Klingenberg C, Fratter C, Carr LJ, Prabhakar P, Kumaraguru N, Gissen P, Cross JH, Jacques TS, Taanman JW, Bindoff LA, Rahman S. The clinical spectrum and natural history of early-onset diseases due to DNA polymerase gamma mutations. Genet Med. 2017 May 4. doi: 10.1038/gim.2017.35. [Epub ahead of print] PubMed PMID: 28471437.

11. Wolf NI, Rahman S, Schmitt B, Taanman JW, Duncan AJ, Harting I, Wohlrab G, Ebinger F, Rating D, Bast T. Status epilepticus in children with Alpers' disease caused by POLG1 mutations: EEG and MRI features. Epilepsia. 2009 Jun;50 (6):1596–1607. doi: 10.1111/j. 1528-1167.2008.01877.x. Epub 2008 Nov 19. PubMed PMID: 19054397.

12. Hirano M, Nishigaki Y, Martí R. Mitochondrial neurogastrointestinal encephalomyopathy (MNGIE): A disease of two genomes. Neurologist. 2004 Jan;**10** (1):8–17. Review. PubMed PMID: 14720311.

13. Clarke SL, Bowron A, Gonzalez IL, Groves SJ, Newbury-Ecob R, Clayton N, Martin RP, Tsai-Goodman B, Garratt V, Ashworth M, Bowen VM, McCurdy KR, Damin MK, Spencer CT, Toth MJ, Kelley RI, Steward CG. Barth syndrome. Orphanet J Rare Dis. 2013 Feb 12;**8**:23. doi: 10.1186/1750-1172-8-23. Review. PubMed PMID: 23398819; PubMed Central PMCID: PMC3583704.

14. Haghighi A, Haack TB, Atiq M, Mottaghi H, Haghighi-Kakhki H, Bashir RA, Ahting U, Feichtinger RG, Mayr JA, Rötig A, Lebre AS, Klopstock T, Dworschak A, Pulido N, Saeed MA, Saleh-Gohari N, Holzerova E, Chinnery PF, Taylor RW, Prokisch H. Sengers syndrome: Six novel AGK mutations in seven new families and review of the phenotypic and mutational spectrum of 29 patients. Orphanet J Rare Dis. 2014 Aug 20;**9**:119. doi: 10.1186/s13023-014-0119-3. Review. PubMed PMID: 25208612; PubMed Central PMCID: PMC4167147.

15. Broomfield A, Sweeney MG, Woodward CE, Fratter C, Morris AM, Leonard JV, Abulhoul L, Grunewald S, Clayton PT, Hanna MG, Poulton J, Rahman S. Paediatric single mitochondrial DNA deletion disorders: An overlapping spectrum of disease. J Inherit Metab Dis. 2015 May;**38**(3):445-457. doi: 10.1007/s10545-014-9778-4. Epub 2014 Oct 29. PubMed PMID: 25352051; PubMed Central PMCID: PMC4432108.

16. Pitceathly RD, Fassone E, Taanman JW, Sadowski M, Fratter C, Mudanohwo EE, Woodward CE, Sweeney MG, Holton JL, Hanna MG, Rahman S. Kearns-Sayre syndrome caused by defective R1/p53R2 assembly. J Med Genet. 2011 Sep;**48** (9):610–617. doi: 10.1136/jmg.2010.088328. Epub 2011 Mar 4. PubMed PMID: 21378381.

17. DiMauro S, Hirano M. MERRF. 2003 Jun 3 [updated 2015 Jan 29]. In: Pagon RA, Adam MP, Ardinger HH, Wallace SE, Amemiya A, Bean LJH, Bird TD, Dolan CR, Fong CT, Smith RJH, Stephens K, editors. GeneReviews® [Internet]. Seattle (WA): University of Washington, Seattle; 1993–2015. Available from www.ncbi.nlm.nih.gov/books/NBK1520/ PubMed PMID: 20301693.

Chapter 3

Clinical Approach in Adults

Michael J. Keogh, Hannah E. Steele and Patrick F. Chinnery

Clinical Approach in Adults

The initial, and perhaps greatest challenge of mitochondrial medicine is the accurate and timely identification of patients with clinical features suggestive of a mitochondrial disorder. As mitochondrial disorders encompass such a heterogeneous spectrum of clinical disease, they present to a multitude of clinical specialties, each with differing levels of experience in considering the diagnosis. Each specialty also varies markedly in its experience in arranging and undertaking the further investigations necessary to diagnose these conditions.

The aim of this section is to outline an approach to the clinical assessment and investigation of mitochondrial disorders, relevant and appropriate to clinicians across all specialties, in order to both appreciate the potential diagnosis and stratify and strategize the necessary subsequent investigations to reach a diagnosis (Figure 3.1).

Likely Routes of Referral

There are three predominant pathways by which a mitochondrial disorder is diagnosed in adults; (A) "oligosymptomatic" presentations, where a patient is referred with an isolated feature of mitochondrial disease, (B) "syndromic" presentations, where patients are referred with several symptoms suggestive of an underlying mitochondrial syndrome and may reflect a "classical syndrome," and (C) multisystem disorders that do not neatly fit into any category of mitochondrial disease: such patients often have been investigated by a myriad of specialties, and due to the atypical and often widespread nature of the phenotype are considered for referral to a tertiary specialist mitochondrial clinic for evaluation (Figure 3.1).

In all potential clinical pathways, a thorough clinical history and evaluation is of paramount importance in determining both the likely presence of a mitochondrial disorder and the range of organs and systems involved, in order to tailor further investigations and referrals as necessary.

History Taking

Presenting complaint: The first step, as with any clinical history, is a thorough history of the presenting symptom/s, with consideration of non-mitochondrial causes constantly appreciated (Figure 3.2).

Additional symptoms within system: Thereafter, it is important to ask specific further questions to determine the presence of other features suggestive of mitochondrial disease within that system, for example inquiring about episodes suggestive of migraines or an encephalopathy in

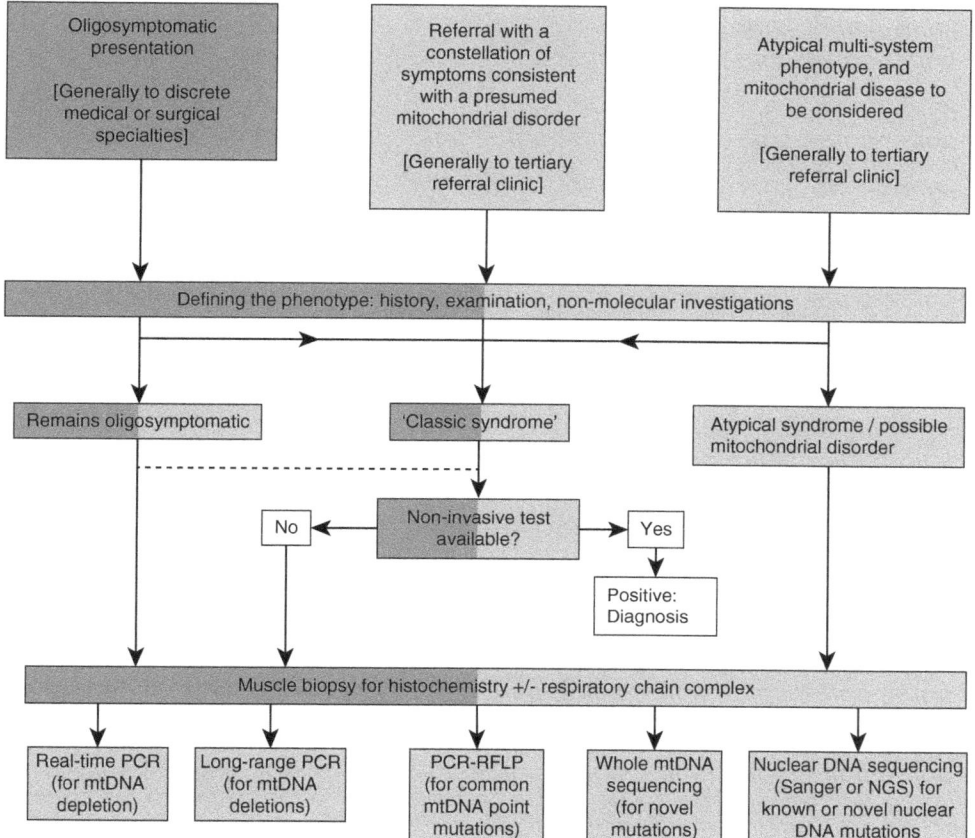

Figure 3.1 A schematic representation of the common clinical and pathological manifestations of mitochondrial disorders in adults

patients presenting with seizures (see Table 3.1); questions that may not be specifically employed in the absence of a suspicion of a mitochondrial disorder (Figure 3.2).

Full-system inquiry: Next, it is important to screen for common features of mitochondrial disease as part of a "full-system inquiry." As discussed in Chapter 1, given that mitochondrial diseases have a particular predilection to affect tissues with a high metabolic demand, the majority of these symptoms are neurological or ophthalmological, but do include general systemic features and symptoms, together with other specific features in other systems. An example set of screening questions is shown in Table 3.1. It is important to remember non-neurological features, particularly evidence of gastrointestinal dysfunction, which is often overlooked.

Past medical history: It is important to take a thorough history of previous medical events, including early life development, something often absent in routine history taking by adult physicians. Specifically, it is prudent to inquire about recurrent maternal miscarriages (suggestive of possible genetic disorders manifesting *in utero*), whether the patient

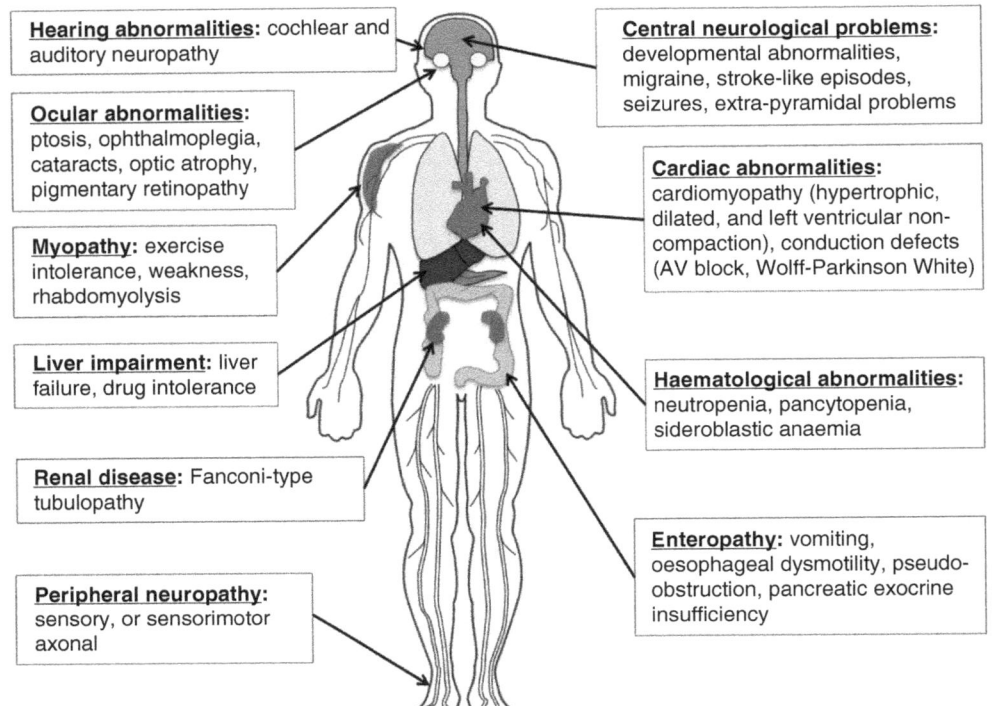

Hearing abnormalities: cochlear and auditory neuropathy

Ocular abnormalities: ptosis, ophthalmoplegia, cataracts, optic atrophy, pigmentary retinopathy

Myopathy: exercise intolerance, weakness, rhabdomyolysis

Liver impairment: liver failure, drug intolerance

Renal disease: Fanconi-type tubulopathy

Peripheral neuropathy: sensory, or sensorimotor axonal

Central neurological problems: developmental abnormalities, migraine, stroke-like episodes, seizures, extra-pyramidal problems

Cardiac abnormalities: cardiomyopathy (hypertrophic, dilated, and left ventricular non-compaction), conduction defects (AV block, Wolff-Parkinson White)

Haematological abnormalities: neutropenia, pancytopenia, sideroblastic anaemia

Enteropathy: vomiting, oesophageal dysmotility, pseudo-obstruction, pancreatic exocrine insufficiency

Figure 3.2 A schematic representation of a clinical investigative pathway for adults with suspected mitochondrial disease
In cases of recognized syndromes or polysymptomatic cases that loosely resemble clinical syndromes, in which a nuclear genetic, homoplasmic mtDNA or heteroplasmic mtDNA point to a pathogenic mutation, then clinical testing of blood with either a candidate gene analysis (or next-generation sequencing, or shortly whole-genome sequencing) should be performed. If these investigations are inconclusive, or if a mtDNA deletion disorder is suspected (e.g., in cases of progressive external ophthalmoplegia), then a muscle biopsy should take place. Thereafter, sequencing of mtDNA can take place as indicated depending on ongoing biochemical findings.

achieved normal developmental milestones, the patient's stature compared to their parents and siblings (i.e., inappropriately short stature), and the patient's sporting and academic ability at school (to inquire about possible subtle early-onset elements of the phenotype).

Some specific questions that are also worth including are to inquire about evidence of transient hepatic or metabolic disturbance in infancy, which could be regarded as irrelevant by adult patients, but are common presentations of mitochondrial cytopathies (Chapter 14).

Social history: Within the social history, it is important to determine the degree of exercise tolerance (suggestive of a mild metabolic or myopathic process), smoking status (potential relevance for LHON (see Chapter 9) and driving status (especially if seizures or cardiac dysfunction are possible within the phenotype) (see Chapter 7).

Drug history: When enquiring about allergies and current medications, it is important to ask about any medications that may exacerbate symptoms for patients with mitochondrial disease (e.g., statins in patients with myopathy).

Table 3.1 Some suggested questions to include within the medical history to try and capture salient information that may assist in determining the likelihood and nature of a mitochondrial disorder

Element of History	Suggested Additional Questions
Presenting complaint	Ensure that non-mitochondrial causes of the presenting symptoms, or common "mimics" are inquired about
Additional "mitochondrial" features within presenting system	Specifically ask about the presence of additional common mitochondrial features in the presenting system/organ (Table 3.2)
Full system inquiry	"Have you ever had any seizures, blackouts, faints or 'funny turns'?" "Have you ever been tested for diabetes?" "Do you have any problems with hearing?" and "Have you had a recent hearing test?" "Do you experience double vision?" "Have you noticed any change in vision including changes in color and clarity?" "Do you ever have palpitations?"
Past medical history	"Are you aware whether your parents had any failed pregnancies before or after your birth?" "Do you know if you were born full term, or had any problems in early life?" "How did you perform academically and in sport at school? How did your siblings perform?" "As a child did you ever have any fits, faints or 'funny turns'?" "Did you have any problems with your liver or require ventilation after birth?" "Has anyone in the family ever died suddenly and unexpectedly?"
Social history	"What sort of exercise can and do you perform now?" "Do you drive?" and "Do you smoke?"
Drug history	"Have you ever had a general anesthetic? Were there any problems?" "Have you ever had any unusual intolerances to medications?"
Family history	A thorough family history should be also taken. It is important to ask about other mitochondrial symptoms from all family members, e.g., in a patient presenting with CPEO as about diabetes and deafness in family members in addition to ocular problems. "Has anyone ever died suddenly and unexpectedly at a young age in the family?"

Family history: With any presentation of a potential genetic disorder, it is of paramount importance to gain an accurate family history. However, it is important not to place too much emphasis on any pedigree information as the clinical features should dominate. Mitochondrial diseases as discussed in Chapters 1 and 5 can be either primary mitochondrial DNA disorders (i.e., as the result of a mutation only within mtDNA) and thus have exclusively maternal inheritance, or they can result from mutations in nuclear DNA and hence be inherited in an autosomal recessive, autosomal dominant of

X-linked fashion. Vitally, one third of patients with a mitochondrial disease has a sporadic disorder and hence a lack of other affected family members certainly should never exclude the diagnosis [1].

If a diagnosis of mtDNA disease is suspected (especially if it may be a primary mtDNA disorder, i.e. there is no male–male transmission in the pedigree), then it is particularly important to ask about subtle clinical features in additional family members, due to the concept of heteroplasmy [2] (Chapter 5). Depending on the mutant to wild-type ratio or "load" of mutant mitochondrial DNA (Chapter 5), then it is common that clinical features can be more or less apparent from organ to organ within a patient, and from member to member within a family, resulting in high clinical heterogeneity. Therefore, it is important to note whether any of the salient features of mitochondrial disorders (Table 3.2) are present in any appropriate family members, not only the patient (or "proband").

Determining pedigree data is also vital to provide genetic counseling to other family members.

Table 3.2 A table of the major clinical features of mitochondrial diseases

System	Feature	Additional Information	Common Syndromes
General	Short stature	Complex and multifactorial etiology reflecting higher levels of heteroplasmy in primary mtDNA disorders together with endocrine dysfunction	MELAS, POLG, KSS, MERRF, NARP, Leigh
	Low muscle mass	A common feature in patients with a coexistent myopathy or growth impairment	POLG, MELAS, MERRF, KSS, NARP, Leigh
	Exercise intolerance	Common presentation of numerous mitochondrial disorders	MELAS, POLG, KSS, NARP, Leigh
Neurological	Ataxia	Variable severity and temporal relationship with disease	KSS, NARP, POLG, MELAS, MERRF
	Dementia	Generally subcortical, but highly variable in severity and pattern	KSS, Leigh, MELAS, MERRF
	Encephalopathy	Often intermittent, but episodes can be extremely prolonged and difficult to manage	MELAS, Leigh, MNGIE, MERRF, KSS
	Myoclonus	Subtle to gross, predominantly cortical myoclonus	MERRF
	Myopathy	Can be ocular, proximal, distal or generalized	POLG, CPEO+, KSS, MERRF

Table 3.2 (cont.)

System	Feature	Additional Information	Common Syndromes
	Peripheral neuropathy	Usually axonal, but can be sensory, motor or mixed	MNGIE, NARP, MERRF, OPA1, KSS
	Seizures	Often progressive, focal or generalized and difficult to control	MERRF, MELAS, Leigh
	SNHL	Often occurs in younger patients, often missed as part of the phenotype by adult physicians	MELAS, MERRF, KSS
	Stroke-like episodes	Predominantly occipital, but not invariably	MELAS
Ophthalmological	CPEO	Can occur in isolation, or as part of a wider syndrome	CPEO, CPEO+, KSS, MELAS, OPA1
	Optic atrophy	May be clinically mild and detected only by an ophthalmologist	LHON, OPA1, MERRF
	Pigmentary retinopathy	Ranges from retinitis pigmentosa to "salt and pepper" fundi	KSS, MELAS, NARP
	Ptosis	Can occur in isolation or as part of CPEO and additional neurological or myopathic features	POLG, KSS, MERRF
Cardiological	Hypertrophic cardiomyopathy	Can occur in isolation or in conjunction with a conduction deficit	KSS, MELAS, MERRF, multiple deletion syndromes
	Conduction defect	Second-degree, third-degree or complete heart block may all be observed	KSS, POLG
	Cardiomyopathy with lactic acidosis	Rare, and generally observed as a severe disorder in infancy	MTO1
	WPW	Rare, but symptoms of palpitations or collapse should raise the possibility	MELAS, MIMyca
Endocrine	Diabetes	Typically presents insidiously in mid-adult life mimicking Type 2 diabetes mellitus	MIDD, MELAS, KSS,
	Hypoparathyroidism	A rare feature, often in severely affected patients with multisystem disease	KSS

Table 3.2 (cont.)

System	Feature	Additional Information	Common Syndromes
	Hypogonadism	Range from primary testicular failure to ovarian dysgenesis. Rare.	POLG, KSS, MELAS
	GH deficiency	Relatively rare, but should always be considered and tested for in patients with appropriate features	KSS, MIDD, MELAS
Gastrointestinal	Profound dysmotility	Rare symptom, but may result in intestinal failure and require nutritional support	MNGIE, Leigh
	Pseudo-obstruction	Can be difficult to manage and exacerbate coexistent symptoms like encephalopathy	MNGIE, MELAS
	Cyclical vomiting	Relatively rare, but should prompt the clinician to consider mitochondrial disease	MELAS
	Hepatic failure	Often early onset, severe and may be fatal	SCO1, POLG, BCS1L, multiple deletion syndromes
Renal	Renal tubular disease	Most commonly a proximal tubular defect	COX10, MELAS, RRM2B
	De Toni-Debre-Fanconi syndrome	One-third of patients with a tubulopathy will be severe enough to fulfill this diagnostic criteria	COX10, MELAS, RRM2B

Additional information expands upon the phenotypic features and the common syndromes in which the features are observed are shown. Key: MELAS – mitochondrial encephalopathy with lactic acidosis and stroke-like episodes, KSS – Kearns-Sayre syndrome, MERRF – mitochondrial encephalopathy with ragged red fibers, NARP – neuropathy ataxia and retinitis pigmentosa, CPEO – chronic progressive external ophthalmoplegia, MNGIE – mitochondrial neurogastrointestinal encephalopathy, LHON – Leber hereditary optic neuropathy, MIDD – maternally inherited diabetes and deafness, POLG – Polymerase gamma

Examination

It is essential to perform a comprehensive and broad "multisystem" examination of patients with suspected mitochondrial disorders (Figure 3.1). Here we provide an overview. A more comprehensive discussion of some of the specialty specific examinations and more rare clinical features are discussed within the relevant specialty specific chapters (Chapters 7–16).

General examination: Ideally an accurate height and weight for the patient should be obtained in clinic, and it is often useful to compare this to relatives accompanying them.

Neurological and Ophthalmological

Higher mental function: Gross assessment of higher mental function, again with relevance to previous premorbid intellectual function and previous educational attainment, is important, primarily to determine the presence of developmental delay, an encephalopathy, or gross deficits in speech (resulting from possible stroke-like episodes for example).

Cranial nerves (II–VI): First, inspection of the eyes and face is of paramount importance to determine the presence of ptosis or any form of abnormal eye alignment in the primary position of gaze, which could suggest CPEO (one of the most common features of mitochondrial disease [3]). Subsequently, assessment of acuity, color perception, ophthalmoscopy and visual fields is vital. This will assist primarily in detecting the presence of optic atrophy (seen in several mitochondrial disorders, see Chapter 9), a pigmentary retinopathy (for example, as seen in Kearns-Sayre syndrome), peripheral visual field defects (from occipital lesions in MELAS) and central field dysfunction (as seen in Leber's hereditary optic neuropathy, for example) (Table 3.2).

Testing ocular movements is not only useful to look for CPEO and other forms of gaze restriction, but will also detect abnormal saccadic and pursuit eye movements that can occur due to cerebellar dysfunction, which is also seen in many mitochondrial disorders with central nervous system involvement.

Cranial nerves (VII–XII): It is important to pay particular attention in the assessment of hearing given the prevalence of sensorineural hearing loss in mitochondrial disease [4]. Any suggestion of hearing impairment from either the history or examination should prompt referral for formal audiological assessment (Chapter 10). Lower cranial nerves are generally not affected with most mitochondrial disorders.

Remainder of Neurological Examination

• **General inspection:** It is important to look for features of reduced muscle bulk, which is commonly observed in mitochondrial myopathies. This is often the only clue to the presence of a mitochondrial myopathy as many patients exhibit normal power on testing. Extra pyramidal movements at rest (e.g., a tremor or chorea) are rare, but can occur in some mitochondrial conditions (Chapter 7).

• **Tone:** This is generally normal in most mitochondrial disorders, and certainly pyramidal tract dysfunction and spasticity is rare, but can be present. Lower motor neuron dysfunction due to peripheral nerve disorders is more common, and often present alongside other supportive morphological features such as pes planus or pes cavus (Chapter 8).

• **Power:** Can be reduced due to a myopathy or lower motor neuron damage from a peripheral neuropathy. Often both features can occur together, causing atypical patterns of weakness and muscle wasting challenging to even the most experienced neurologist.

• **Reflexes:** Are generally normal in most patients without peripheral nerve involvement. Hyperreflexia is rare due to the lack of pyramidal involvement in most mitochondrial disorders.

• **Coordination:** Cerebellar dysfunction is common in many mitochondrial disorders, so specific assessment of gait and limb movements are important.

Cardiological

A standard assessment of the cardiovascular system should be undertaken in clinic. Blood pressure is often normal or low in the presence of a cardiomyopathy. Heart rate may be normal, or can be reduced in the presence of heart block, which occurs in several mitochondrial disorders (e.g., Kearns-Sayre syndrome) (Chapter 11).

Palpation of the precordium may reveal a right ventricular heave in the context of an underlying cardiomyopathy, but auscultation is invariably normal as valvular heart disease is not known to be a feature of any mitochondrial disorder.

Features of heart failure (raised jugular venous pulse, pedal edema) should also be sought, but reflect more advanced disease and are unlikely to be found incidentally in patients presenting with non-cardiological symptoms.

Endocrine

While no specific formal examination per se of the endocrine system is performed, it is important to look for features of endocrine dysfunction throughout the consultation and examination. In our practice, as part of the neurological examination when assessing for muscle bulk, it is important to note whether the patient is generally underweight, which is a common feature of diabetes in patients with mitochondrial disorders and in contrast to the more common Type 2 diabetes within the population [5]. It is also important to look for features of growth hormone deficiency (short stature – though this is not the sole cause in many mitochondrial patients) and for features of hypogonadism, such as a lack of body hair and gynecomastia (Chapter 13).

Clinical features consistent with hypothyroidism, hypoadrenalism or autoimmune endocrine disorders are generally not associated with mitochondrial disease but should always be considered.

Gastrointestinal and Respiratory

In the absence of any specific elements of the history to suggest respiratory or gastrointestinal disease, these examinations are likely to be normal in most mitochondrial diseases, though exceptions do occur, and are discussed in Chapters 12 and 14.

Investigations

In clinic: If a mitochondrial disorder is suspected, even with an oligosymptomatic presentation, then an ECG, urine dipstick (to look for either glucose or proteinuria), and a blood test for serum blood sugar and HbA1c are mandatory in clinic. Normally, additional blood tests are also taken at this time, in particular creatinine kinase (CK) to highlight any suggestion of muscle breakdown.

Other blood tests may also be taken at baseline, though this is not universal and may not be helpful; serum urea and electrolytes may be helpful in the investigation of renal dysfunction, and liver function testing for hepatic disease (though more common in neonates than adults). In the absence of any other specific features, additional biochemical blood tests are rarely helpful in adults present in the clinic. High lactate levels are uncommon in adults with mitochondrial disease who are not acutely unwell.

Noninvasive Outpatient Investigations

Routinely Performed

Echocardiogram: An echocardiogram is usually performed in all patients at baseline given the asymptomatic nature of many cardiomyopathies and the significant associated risk of mortality and morbidity if present.

Formal eye assessment: This is usually performed by an experienced ophthalmologist who can detect subtle features of optic atrophy and the retinal abnormalities exhibited in mitochondrial disease (Table 3.3).

Table 3.3 A table of useful investigations and their possible outcomes in the investigation of common features of mitochondrial disorders

Symptom	Investigation	Possible findings	Possible treatment/ management
Seizures/ encephalopathy	EEG	Epileptiform activity, acute slowing	Anticonvulsants
Stroke-like episodes	MRI brain	High T2 signal with posterior predominance and inconsistent with established vascular territories on MRI	No established therapy
	EEG brain	EEG to detect the presence of any ongoing ictal activity that should be treated aggressively	Aggressive acute management of seizure activity
Myopathy	CK	Normal (where mild myopathy or low muscle mass), high (moderate – severe myopathy), very high (CoQ_{10} deficiency)	
	EMG	Consistent with a myopathy +/- coexistent neuropathy	
Sensory or sensorimotor neuropathy	Nerve conduction studies	Axonal sensory or sensorimotor neuropathy	Symptomatic therapy
Ocular symptoms	Ophthalmological assessment	Oculomotor abnormalities, optic atrophy, pigmentary retinopathy	Corrective lenses, surgery, prisms, arrange for visual aids at home
Hearing loss	Audiography	Sensorineural hearing loss	Hearing aids, cochlear implantation

Table 3.3 (cont.)

Symptom	Investigation	Possible findings	Possible treatment/ management
Dysphagia	Videofluoroscopy or swallowing studies	Exclude structural lesion, pharyngeal or esophageal dysmotility	Speech and language therapy input, aspiration pneumonia awareness
Cardiological abnormalities	ECG, echocardiogram, CMR	Cardiomyopathy, conduction abnormality	Antiarhythmics, ACE inhibitors, ARBs, PPM, ICD
Respiratory weakness	Pulmonary function studies, sleep studies	Decreased FEV1, FVC and SNIPs	CPAP, BiPAP
Additional investigations	Serum lactate	Normal or elevated	
	CSF protein and biochemistry	Normal or elevated lactate, raised protein	
	CT brain	Basal ganglia calcification	
	MRI brain	Basal ganglia signal abnormalities, nonspecific white matter abnormalities, stroke-like lesions, cerebellar or brain-stem atrophy, or normal. MR spectroscopy may demonstrate elevated lactate	

Key: EEG – electroencephalogram, MRI – magnetic resonance imaging, CK – creatinine kinase, EMG – electromyogram, ECG – electrocardiogram, CMR – cardiac MRI, ACE – angiotensin-converting enzyme, ARM – angiotensin II receptor blocker, PPM – permanent pacemaker, ICD – implantable cardioverter defibrillator, CSF – cerebrospinal fluid, CT – computed tomography, FEV – forced expiratory volume, FVC – forced vital capacity, SNIPs – nasal sniff inspiratory pressures, CPAP – continuous positive airway pressure, BiPAP – bi-level positive airway pressure

Commonly Performed

Neurological

MRI imaging of the brain: This is generally undertaken only when patients describe clear neurological symptoms, and is not considered a useful screening test for mitochondrial disease. MRI imaging is best for highlighting cortical and most of the subcortical abnormalities associated with mitochondrial disease. CT scanning is, however, better for detecting calcium deposition seen within the basal ganglia and present in a small number of mitochondrial disorders (Chapter 7).

EEG: This is a particularly useful investigation to determine the presence of ongoing subclinical ictal activity in disorders such as MELAS (Chapter 7). However, in the absence of clinical symptomatology suggestive of ongoing ictal events, an EEG is likely to be useful only to support a diagnosis of a diffuse encephalopathy, but is highly unlikely to assist in defining the cause.

Nerve conduction studies and electromyogram: These studies are particularly useful to confirm and define the extent of myopathy and neuropathic processes. These features are seen in a multitude of disorders and are often subclinical in nature (Chapter 8).

Cardiological

Apart from an ECG and echocardiogram, further investigations are usually not required within the initial investigative process. Investigations such as prolonged ECG recording are rarely helpful as most conductive defects that result from mitochondrial disease are seen in ambulatory ECGs. Autonomic function testing is rarely helpful given the limited involvement of this system in mitochondrial disease.

Endocrinological

In the presence of symptoms or features suggestive of endocrine dysfunction, full screening of the hypothalamo-pituitary adrenal axis (parathyroid hormone [PTH], growth hormone [GH], calcium, phosphate, adrenocorticotropic hormone [ACTH], thyroid-stimulating hormone [TSH], luteinizing hormone [LH], follicle-stimulating hormone [FSH], prolactin [PRL] and testosterone) should be checked given that mitochondrial disorders have a predilection to cause anterior pituitary dysfunction over other elements of the endocrine system. While this does not hugely assist in narrowing down the diagnosis given the rarity of endocrinological involvement in mitochondrial disease (Chapter 13), it obviously has immediate and profound treatment implications for the patient. As mentioned earlier, it is essential to check a blood glucose and HbA1c because diabetes mellitus is a common feature of mitochondrial disease.

Molecular Genetic Investigations

Once the full phenotype has been compiled from the clinical history, examination and data from the aforementioned investigations (when appropriate), then proceeding to molecular testing becomes possible (Figure 3.2).

It is important to note that molecular testing in mitochondrial disorders can range from the extremely simple to the extraordinarily complex. The specific nature of the investigations and further information relating to the molecular diagnostics platforms are explored in greater detail in Chapter 4.

The aim of this book, and the rationale for the multitude of clinical cases compiled within each chapter from herein, is to enable the reader to "feel" his or her way through approaching and arranging the differing diagnostic processes, which is one of the hardest facets of mitochondrial medicine.

General Rules

- If the case is consistent with a defined, discrete clinical syndrome, and if a noninvasive test is available for that syndrome (e.g., blood or urine analysis for the m.3243 A>G mutation in suspected MELAS), then that should be tested first.
- In oligosymptomatic cases, then noninvasive tests for a similar "classical syndrome" can be performed first, for example, in a case of young man with an occipital stroke and

previous sensorineural deafness, then noninvasive tests for the m.3243 A>G mutation for MELAS (the most similar "classical syndrome" and genotype) can be arranged. However, if this is negative, then this does not preclude the diagnosis of MELAS given that both (a) the m.3243 A>G mutation could be missed due to low levels of heteroplasmy of the mutation in blood despite having high levels of heteroplasmy in brain, and (b) this test would not exclude other mtDNA or nuclear DNA mutations causing a MELAS phenotype. It is likely in coming years that next-generation sequencing will be able to screen quickly the entire mitochondrial genome for all known homoplasmic and heteroplasmic variants (even low-level) within the blood or urine, together with all nuclear genes if required (through whole genome sequencing [WGS]). This, it is hoped, might negate the need to proceed to muscle biopsy testing for many patients, offering the ability to screen for mitochondrial and nuclear DNA mutations causing mitochondrial disease. The caveats to this approach are that low-level heteroplasmic mtDNA point mutations may be missed (if even present in blood) and mtDNA deletions are at present extremely difficult to detect with existing next-generation sequencing platforms.

- If either a noninvasive test is negative, or a noninvasive test is not appropriate, then a muscle biopsy should be performed.
- If a muscle biopsy also supports mitochondrial dysfunction, then the nature of the dysfunction will help guide the next most appropriate investigations, ranging from screening for mtDNA mutations (point mutations, deletions or depletion) or for nuclear gene defects (Chapter 4).

Figure 3.1 demonstrates a schematic of how this approach fits within clinical practice and referral paradigms.

Specific Syndromes

There are several "classical syndromes" that commonly present in adults, and importantly, there is an even greater number of patients with mitochondrial disorders that are diagnosed in infancy and are subsequently managed by adult physicians. In what follows, we discuss two common cases of both primary and secondary mitochondrial disorders seen in adult clinical practice and review their salient clinical features and the investigative approach to determine the molecular diagnosis.

A summary of the major features of all common mitochondrial syndromes is described in Table 3.4. In addition, "case-based" presentations and discussions of all these conditions are featured throughout Section 2 of this book.

CPEO

The most frequent syndrome presenting in adults is CPEO. It can develop at any age, though generally in mid-adult life, and presents with progressive, often asymmetrical ptosis in conjunction with a progressive extranuclear ophthalmoplegia.

Patients also often have a coexistent and predominantly proximal skeletal myopathy. Importantly, patients may also have additional features of the more severe Kearns-Sayre syndrome (KSS) such as a cardiac conduction block, cerebellar ataxia, "salt and pepper" retinopathy and, occasionally, diabetes or sensorineural deafness. Patients with these additional features are termed "CPEO plus," because they neither have isolated CPEO nor do they do not fulfill the diagnostic criteria for KSS [6]. The reason these disorders lie on the

Table 3.4 A table of the major mitochondrial disease syndromes

Syndrome	Primary Features	Secondary Features	Inheritance	Common Mutations
CPEO	Progressive external ophthalmoplegia with bilateral ptosis	Mild facial or proximal myopathy	S/M	95 percent sporadic; generally caused by mtDNA deletions; more than 150 described
POLG	Sensory or cerebellar ataxia	Highly variable: ptosis, PEO, myopathy, myoclonus, epilepsy (continuum of conditions previously felt to be discrete diagnostic entities, e.g., Alpers' syndrome, MEMSA and arPEO or adPEO)	AR/AD	Numerous pathogenic SNPs described in POLG
Kearns-Sayre	CPEO with onset before 20 years old, with pigmentary retinopathy plus one of; CSF protein > 1 g/l, cerebellar ataxia or heart block	Dementia, sensorineural deafness (50 percent), diabetes, myopathy, hypoparathyroidism and dysphagia	S/M	Generally sporadic mtDNA deletions; more than 100 described
MELAS	Stroke-like episodes (typically before 40 years of age); encephalopathy with seizures and possibly dementia; mitochondrial myopathy	Recurrent headaches, recurrent vomiting, sensorineural deafness	M	3243A>G 3271 T>C 3251A>G
MERRF	Myoclonus (generally cortical), seizures, encephalopathy and ataxia	Dementia, peripheral neuropathy, cervical lipomatosis, sensorineural deafness and optic atrophy	M	8334A>G 8256 T>C
NARP	Pigmentary retinopathy, ataxia and peripheral neuropathy	Dementia, SNHL, short stature, seizures, basal ganglia MRI changes and clinical overlap with Leigh's syndrome	M	8993 T>G/C

Table 3.4 (cont.)

Syndrome	Primary Features	Secondary Features	Inheritance	Common Mutations
MNGIE	Gastrointestinal dysmotility, cachexia, peripheral neuropathy, ptosis and SNHL	GORD, vomiting, distal weakness and leukoencephalopathy on MRI	AR	Several mutations in TYMP described
OPA1	Painless visual loss	SNHL, peripheral neuropathy and ataxia	AD	More than 200 mutations in OPA1 described
LHON	Subacute painless bilateral visual failure (generally in men [ratio 4:1 with women])	Dystonia and cardiac arrhythmia	M	11778 G>A 14484 T>C 3460 G>A

Each syndrome encompasses many of the key features of mitochondrial disorders seen in Table 3.2.
Key: AD – autosomal dominant, AR – autosomal recessive, CPEO – chronic progressive eternal ophthalmoplegia, CSF – cerebrospinal fluid, GORD – gastroesophageal reflux disease, LHON – Leber's hereditary optic neuropathy, MERRF – mitochondrial encephalopathy with ragged red fibers, MELAS – mitochondrial encephalomyopathy with lactic acidosis and stroke-like episodes, MEMSA – myoclonic epilepsy myopathy sensory ataxia, MNGIE – mitochondrial neurogastrointestinal encephalopathy, MRI – magnetic resonance imaging, NARP – neuropathy ataxia retinitis pigmentosa, OPA1 – optic atrophy type 1, PEO – progressive external ophthalmoplegia, POLG – polymerase gamma, S – sporadic, SNHL – sensorineural hearing loss, TYMP – thymidine phosphorylase.

same clinical spectrum is that they generally share a common genetic etiology (mtDNA deletion disorders [discussed later]).

As discussed in the previous section, an ECG, echocardiogram and serum testing for diabetes should be performed mandatorily for patients presenting with CPEO, and such tests alongside clinical assessment will stratify the phenotype between CPEO and CPEO plus.

There are a multitude of genetic causes of CPEO, but the most common is a sporadic single mtDNA deletion. The deletion is generally not detectable in blood but present either exclusively or in much higher levels of heteroplasmy in skeletal muscle [7]. Therefore, a muscle biopsy is generally the most appropriate first step for molecular investigation in this context.

The results of the muscle biopsy in CPEO typically show ragged red fibers (RRF) with hyperactive succinate dehydrogenase (SDH) staining, and additional mosiac cytochrome c oxidase (COX) defects (Chapter 4). In such a clinical phenotype in conjunction with these muscle biopsy results, long-range PCR is likely to find the suspected single large-scale DNA deletion. Further cases of CPEO are presented in greater detail in Chapters 7 and 9.

DNA Polymerase-γ (POLG) Disease

Adult-onset POLG diseases are an overlapping spectrum of disorders that were previously assumed to be etiologically discrete, but that all result from mutations in the nuclear DNA-encoded gene *POLG*. The majority of pathogenic mutations are homozygous or compound heterozygous, though autosomal dominant forms of disease can occur.

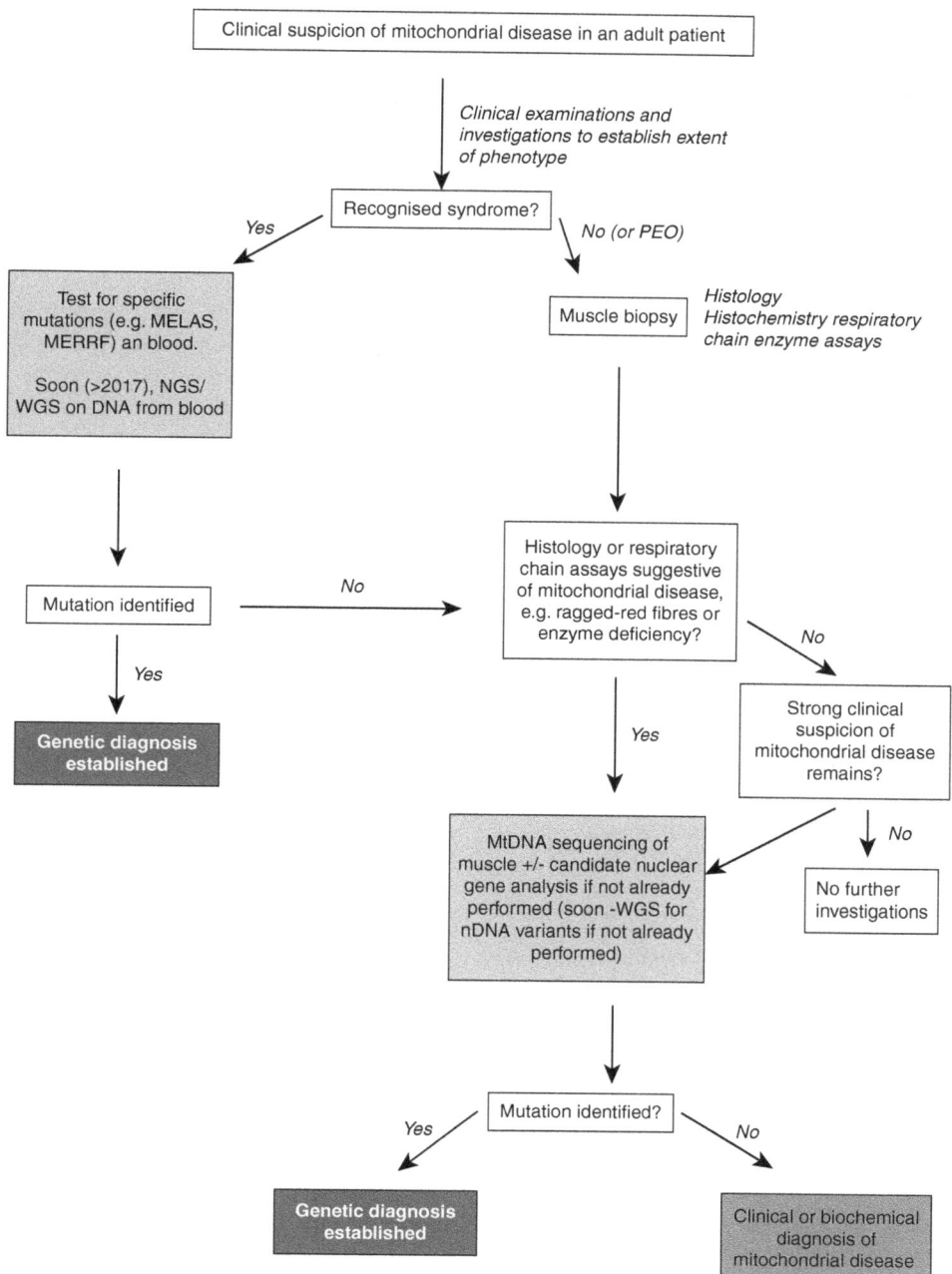

Figure 3.3 A schematic representation of the investigation of how the investigation of potential mitochondria disorders is likely to evolve in the near future.
Key: PEO – progressive external ophthalmoplegia, NGS – next-generation sequencing, WGS – whole genome sequencing

While no discrete diagnostic criteria exist, the major autosomal recessive *POLG* syndromes in adults can broadly be grouped into (a) MEMSA (myoclonic epilepsy myopathy and sensory ataxia); (b) ANS (ataxia neuropathy spectrum), which is a broad cohort of patients presenting with either central or peripheral ataxia often with associated epilepsy and encephalopathy; and (c) arPEO (autosomal recessive PEO), in which PEO is often present in association with depression, parkinsonism and premature ovarian failure. The major autosomal dominant syndrome, adPEO, presents similarly to arPEO. It is important to note that when *POLG* syndromes occur in children, the phenotype can vary markedly (causing conditions such as Alpers Huttenlocher), which involves profound liver disease (see Chapter 14 for further information).

Given the breadth of clinical symptomatology in adults, again an ECG, echocardiogram, Hba1c, together with an EEG, MRI brain and nerve conduction studies due to the preponderance of neurological symptoms are likely to be performed on the majority of patients.

Often, patients with phenotypes mimicking other "classical syndromes," e.g., MEMSA, which mimics mitochondrial encephalopathy with ragged red fibers (MERRF) to some degree, will have noninvasive investigations performed for that condition (e.g., the m.8344 A>G mutation – the most common MERRF genotype). Thereafter, a muscle biopsy will be required; however, they can give numerous different findings in *POLG* disorders ranging from discrete biochemical abnormalities to depletion of mitochondrial DNA or multiple DNA deletions (in contrast to a single deletion seen in CPEO). Such findings will often result in a series of screens of both nuclear and mitochondrial DNA from muscle tissue, including looking for germline *POLG* mutations, in order to reach the diagnosis. Again, the advent of next-generation clinical sequencing that can screen the whole nuclear genome in addition to the mitochondrial genome is likely to facilitate the diagnosis of nuclear mutations causing mitochondrial disorders significantly in the near future and may result in a significant reordering of the diagnostic algorithm shown in Figure 3.1 to that seen in Figure 3.3.

References

1. Chinnery PF. Inheritance of mitochondrial disorders. *Mitochondrion* 2002; 2(1–2): 149–155.

2. Cree LM, Samuels DC, Chinnery PF. The inheritance of pathogenic mitochondrial DNA mutations. *Biochimica et biophysica acta* 2009; 1792(12): 1097–1102.

3. McFarland R, Taylor RW, Turnbull DM. The neurology of mitochondrial DNA disease. *The Lancet Neurology* 2002; 1(6): 343–351.

4. Kullar PJ, Quail J, Lindsey P, et al. Both mitochondrial DNA and mitonuclear gene mutations cause hearing loss through cochlear dysfunction. *Brain: A Journal of Neurology* 2016; 139(Pt. 6): e33.

5. Schaefer AM, Walker M, Turnbull DM, Taylor RW. Endocrine disorders in mitochondrial disease. *Molecular and Cellular Endocrinology* 2013; 379(1–2): 2–11.

6. DiMauro S, Hirano M. Mitochondrial DNA deletion syndromes. In Pagon RA, Adam MP, Ardinger HH, et al., eds. *GeneReviews*(R). Seattle (WA); 2011.

7. Taylor RW, Schaefer AM, Barron MJ, McFarland R, Turnbull DM. The diagnosis of mitochondrial muscle disease. *Neuromuscular Disorders: NMD* 2004; 14(4): 237–245.

Laboratory Investigation of Mitochondrial Diseases

Robert W. Taylor and David R. Thorburn

Introduction

A multidisciplinary approach to the diagnosis of mitochondrial disease is often used to piece together information from clinical phenotyping and investigations, histopathological and biochemical testing to help refine molecular genetic screening as guided by traditional diagnostic algorithms for mitochondrial disease [1, 2], although more and more frequently the use of next-generation sequencing strategies (either target-capture using a bespoke gene panel, whole exome or whole genome sequencing) is being introduced at earlier points of the diagnostic pathway. In any case, the availability of a range of functional testing, including assessment of individual respiratory chain enzyme activities in a tissue biopsy or cellular respiration in a patient cell line, remains useful to help validate candidate pathogenic variants. While not described in detail here, it is important to note that analysis of metabolites or biomarkers in urine, blood or cerebrospinal fluid remains important for excluding some differential diagnoses and providing support for specific candidate genes, as described in Table 4.1. In this chapter, we briefly summarize the three main disciplines used in the laboratory investigation of suspected mitochondrial disease – tissue histopathology, biochemical assessment of mitochondrial function and molecular genetic analyses.

Table 4.1 Metabolite and biomarker analysis in diagnosis of mitochondrial diseases

Markers of Mitochondrial Disease

Elevated plasma FGF21, GDF15, lactate or alanine levels are useful markers that can increase suspicion of mitochondrial disease, but they can be increased secondarily and are not highly sensitive.[64–67]

Markers to Exclude Differential Diagnoses

Mass spectrometric and other analyses of organic acids, amino acids and acylcarnitines can exclude differential diagnoses such as organic acidurias, amino acidurias and fatty acid oxidation disorders.[68]

Markers Supportive of Specific Genetic Disorders

3-methylglutaconic aciduria in *TAZ, TMEM70, SERAC1, OPA3, DNAJC19, HTRA2* defects[69, 70]
Methylmalonic aciduria in *SUCLA2, SUCLG1* defects[71]
Ethylmalonic aciduria in *ETHE1*[72]
Increased plasma thymidine and deoxyuridine in *TYMP* defects[73]
2,3-dihydroxy-2-methylbutyricaciduria in *HIBCH* and *ECHS1* defects[74]
nonketotic hyperglycinemia in *LIAS, NFU1, BOLA3, IBA57, GLRX5* defects[75]

Histopathology

Laboratory-based diagnosis has long relied upon the analysis of a clinically relevant tissue, often post-mitotic skeletal muscle, as the affected organ or surrogate, but other tissues (e.g., cardiac muscle and liver) may be appropriate in certain circumstances. In many patients, histopathological hallmarks are present – both histological and histochemical changes – that are indicative of mitochondrial respiratory chain dysfunction, although the finding of a normal biopsy does not necessarily preclude the diagnosis of mitochondrial disease.

Hematoxylin and eosin (H&E) and modified Gomori trichrome are histological stains that are widely used to assess basic muscle morphology, providing information on fiber size and the presence of any abnormal inclusions or central nuclei that may be indicative of muscle denervation or other pathology (Figure 4.1A). The modified Gomori trichrome stain specifically allows the detection of ragged red fibers (RRF), characterized by a "fiber cracking" appearance and abnormal subsarcolemmal proliferation of mitochondria resulting from a compensatory response to a respiratory chain biochemical defect. RRF are rarely seen in children and can be associated with normal activities of oxidative enzyme activities. A key finding in patients with either primary mtDNA disease or nuclear-driven mitochondrial disease affecting the expression of the mitochondrial genome is the demonstration of respiratory-deficient muscle fibers, in particular those lacking histocytochemical cytochrome c oxidase (COX) activity. COX-deficient fibers were first identified in patients with chronic progressive external ophthalmoplegia (CPEO) [3] as demonstrated by a loss of the oxidized 3,3'-diaminobenzidine tetrahydrochloride (DAB) reaction product used in the histochemical assay (Figure 4.1A). This can affect all muscle fibers – a global COX defect – and is typically seen in patients with homoplasmic mt-tRNA mutations [4] or Mendelian-inherited disorders of COX assembly [5–7]; it is, however, more often observed as a mosaic pattern of deficiency with a mixture of COX-deficient and COX-positive fibers within the same biopsy sample. The demonstration of a mosaic COX-deficiency can be enhanced by combining the individual COX reaction with the histochemical demonstration of succinate dehydrogenase (SDH) activity in which a blue, nitroblue tetrazolium (NBT) reaction product is formed; SDH, a component of complex II and the TCA cycle, is fully encoded by the nuclear genome, whereas COX has three catalytic subunits encoded by mtDNA. The presence of a mosaic pattern of COX-deficient, SDH-reactive muscle fibers (Figure 4.1A) is therefore considered a pathological hallmark of an mtDNA-related defect [8, 9]. The sequential COX/SDH histochemical reaction has been extensively used to explore the molecular mechanisms involved in mtDNA disease, with COX-deficient fibers showing high levels of mutated mtDNA and lower levels of wild-type mtDNA in patients with heteroplasmic mtDNA defects [10], the role of mitochondrial changes in aging and age-related disease [11, 12], the role of secondary mitochondrial defects in other muscle pathologies [13] and also as a useful tool for lineage tracking of particular stem cell populations in epithelial tissues that accumulate high levels of somatic mtDNA mutations [14].

Although the sequential COX/SDH histochemistry assay is an excellent marker for cells deficient in one component of the mitochondrial respiratory chain, namely complex IV, it will not identify cells deficient in other respiratory chain complex activities. Such defects can be caused by a range of genetic defects, including mtDNA and nuclear mutations affecting the structural components or assembly factors of other complexes or even mtDNA-related abnormalities causing generalized mitochondrial translation defects. Immunohistochemical techniques, using specific monoclonal antibodies to individual subunits of mitochondrial

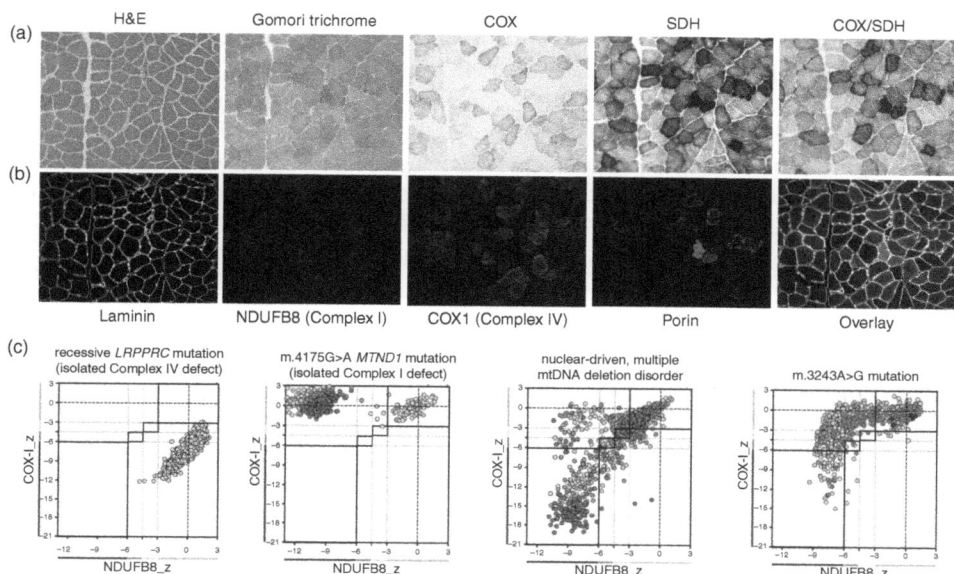

Figure 4.1 Histological, histochemical and immunohistochemical hallmarks of mitochondrial pathology in mitochondrial disease. **A**, serial skeletal muscle (quadriceps) sections from a patient with a nuclear-driven, multiple mtDNA deletion disorder were stained for hematoxylin and eosin (H&E) and modified Gomori trichrome to assess basic muscle morphology and demonstrate the presence of ragged red fibers – a sign of abnormal mitochondrial accumulation – respectively. The individual COX, SDH and sequential COX/SDH histochemical reactions show fibers manifesting mitochondrial accumulation and focal COX-deficiency. **B**, serial sections of the same muscle biopsy sample subjected to a quadruple immunofluorescence assay that can quantitate the expression of complex I (NDUFB8 subunit), complex IV (COXI subunit), laminin (to mark muscle fiber boundaries) and a mitochondrial mass marker (porin), all within a single 10μm section (all images taken at 10X). **C**, mitochondrial respiratory chain expression profile plots (see Rocha et al. [17]) showing COX-I, NDUFB8 and porin protein levels in single muscle fibers from patients with different mitochondrial genetic defects (as indicated), highlighting the different profiles relating to mitochondrial genotype. Each dot represents the measurement from an individual muscle fiber, color-coded according to its mitochondrial mass (very low: blue; low: light blue; normal: light orange; high: orange; very high: red). Thin black dashed lines indicate the SD limits for the classification of fibers, lines next to the x and y axes indicate the levels of NDUFB8 and COX-I, respectively (beige: normal; light beige: intermediate (+); light blue: intermediate (-); blue: deficient). Bold dashed lines indicate the mean expression level of normal fibers. (A black and white version of this figure will appear in some formats. For the color version, please refer to the plate section.)

respiratory chain complexes, have long been used in the study of mitochondrial biochemical defects in tissue sections to gain a comprehensive picture of the extent of respiratory chain deficiency in muscle biopsy samples [15, 16]. Such techniques can provide additional information to that obtained from the *in vitro* measurements of respiratory chain activities in muscle homogenates since in patients with heteroplasmic mtDNA mutations, for example, the biochemical defect can present as a "mosaic" rather than a generalized defect and thus may not be detectable in homogenized tissue samples. As no routine histochemical assays are available to evaluate respiratory chain components other than complex IV (COX) and complex II (SDH), quantitative, quadruple immunofluorescence assays such as the one recently devised by Rocha and colleagues [17] have the potential to enhance the range of tools available to diagnose mitochondrial disease at a biochemical level as well as evaluate the full extent of mitochondrial respiratory deficiency in patient tissue sections at a single fiber level in a range of diagnostic and research-based applications (Figure 4.1B). This technique enables the accurate quantification of the two most commonly affected respiratory chain components, namely complexes I and IV [17], together with a mitochondrial mass marker (porin) in individual muscle fibers on a single 10μm tissue section. The semiautomatic quantification of a large number (>1000) of muscle fibers is undertaken by labeling laminin to define skeletal muscle fiber boundaries, facilitating automated image analysis, which is exclusively based on intensity measurements, allowing plots to be drawn that not only show relative amounts of respiratory chain proteins but relate this to mitochondrial content (based on porin expression levels) (Figure 4.1 C).

Biochemistry

The first patient with a biochemically characterized mitochondrial disease was reported by Luft and colleagues in 1962.[18] Subsequently, mitochondrial biologists and diagnostic scientists have utilized a range of functional techniques to investigate patients with suspected mitochondrial disease for evidence of respiratory chain dysfunction. These typically categorize patients according to their biochemical phenotype – respiratory chain defects involving either single enzyme complexes or multiple respiratory chain components – to streamline the genetic studies that have led to the identification of several hundred mitochondrial disease genes [19].

Diagnostic centers specializing in mitochondrial disorders employ numerous techniques to assess mitochondrial function, including the assessment of individual mitochondrial respiratory chain activities *in vitro* within mitochondrially enriched tissue fractions [20]. Although useful for identifying widespread mitochondrial defects, this technique has some limitations; it requires large quantities of muscle (typically 50–100 mg tissue) and may fail to detect subtle biochemical deficiencies, especially when complicated by mtDNA heteroplasmy. Some groups continue to have access to fresh muscle biopsy samples, permitting the measurement of mitochondrial substrate oxidation, ATP production and respiratory flux in intact mitochondria, but the overwhelming majority now assess frozen muscle specimens to facilitate the transport of samples to specialist centers, often over large distances. Different centers use different substrates and/or electron acceptors in enzyme assays, making universal "normal" ranges almost impossible to define. Consequently, consensus protocols for assessing respiratory chain activities have been difficult to achieve [21], although implemented in some countries [22].

Unfortunately, only the activities of complexes I–IV can be reliably assessed in frozen muscle [20], although the assessment of linked, spectrophotometric assays that measure electron transfer through a portion of the respiratory chain (e.g., succinate-cytochrome c reductase, which comprises complexes II+III) and coenzyme Q_{10} levels is feasible in frozen tissue samples. Measurement of respiratory chain enzymes in cultured cells from patients – often dermal fibroblast cell lines – can be useful to confirm a biochemical defect found in another tissue, although it is increasingly being recognized across many specialist laboratories that ~50 percent of patients with a proven respiratory chain enzyme defect detected in a tissue such as skeletal muscle, liver or heart will express the same defect in cultured skin fibroblast [20]. Nevertheless, access to patient cell lines offers a range of complementary protocols to substantiate a biochemical diagnosis, including the assessment of cellular respiration and oxygen consumption using microscale oxygraphy [23, 24], which can be supplemented with gel-based electrophoretic techniques to assess the stability and expression of mitochondrial proteins or the assembly of respiratory chain components into functional respiratory chain complexes using Blue-Native PAGE [25, 26]. More recently, newer technologies such as metabolic profiling of plasma and urinary analytes [27] and mass spectrometry applications [28] are being utilized to characterize mitochondrial biochemical defects and to identify new players in mitochondrial biology, and may certainly impact the way in which biochemical profiling of mitochondrial dysfunction is performed in the absence of tissue biopsy samples as genetic testing takes a more prominent place within diagnostic algorithms.

Molecular Genetics

Mitochondrial disorders comprise more than 250 monogenic diseases, which can display maternal, autosomal recessive, autosomal dominant or X-linked inheritance [19, 29]. Over the past two decades, molecular genetic diagnosis has typically started with testing a noninvasive sample such as blood or urine for a panel of mtDNA point mutations or deletions. Given the hundreds of potential candidate nuclear genes, Sanger sequencing of single nuclear genes in blood has usually been performed only when the clinical presentation suggested a classic syndrome with a strong genotype/phenotype correlation. The relatively low diagnostic yield of such testing meant that most patients proceeded to biopsy of muscle or other tissues, which could provide a biochemical diagnosis and give a sample for extended mtDNA analyses and/or guide more extensive testing of specific nuclear genes [30]. The introduction of next-generation DNA sequencing (NGS) methods has begun to turn this diagnostic paradigm on its head [31]. It is becoming increasingly feasible to spare patients an invasive biopsy by sequencing mtDNA, large gene panels, whole exomes or whole genomes and only perform a muscle biopsy if those results are ambiguous, requiring functional validation.

Mitochondrial DNA Mutations

Point mutations are the most common type of mtDNA mutation and can be tested for by a number of methods, whose strengths and weaknesses are summarized in Table 4.2. Panels of point mutations allow testing for the most common mutations associated with conditions such as mitochondrial encephalomyopathy lactic acidosis and stroke-like episodes (MELAS), myoclonic epilepsy with ragged red fibers (MERRF), Leber hereditary optic neuropathy (LHON) and Leigh syndrome, using methods such as restriction fragment length polymorphisms, allele-specific hybridization methods or primer extension analyses such as Sequenom. These methods are largely being superseded by NGS methods, which potentially detect all mtDNA point mutations and deletions rather than a subset [32, 33]. However, mutation panels retain an advantage that they can often be performed on DNA of relatively poor quality, as typically found in urine sediment, whereas NGS methods require high-quality, intact mtDNA. This is particularly relevant when testing for the common m.3243A>G mutation associated with MELAS, which can be absent in the blood of adult patients but is retained in urine DNA [34], with urine heteroplasmy being the best predictor of clinical outcome for this mutation [35].

Rearrangements of mtDNA are commonly seen as either single or multiple deletions, which are frequently found in patients with conditions such as CPEO, Kearns-Sayre and Pearson syndrome [36]. mtDNA deletions typically disrupt multiple mtDNA genes encoding respiratory chain subunits and tRNAs. Occasional patients have single duplications of mtDNA that can potentially disrupt expression of genes at the duplication breakpoints,

Table 4.2 Major methods for molecular genetic diagnosis of mitochondrial DNA mutations

Method	Strengths	Weaknesses
Point mutation panel, e.g., by PCR and Sequenom or RFLP test	Simple, quick, LQ	Limited number of mutations tested, semi-quantitative, relative lack of sensitivity
Sanger sequencing	Detects any point mutation, LQ	semi-quantitative, relative lack of sensitivity
Southern blotting	Detects large deletions/insertions	Slow, labor-intensive, HQ
long-range PCR	High sensitivity	Non-quantitative, HQ, could miss large insertions
quantitative PCR	Quantitation of large deletions/insertions	relative lack of sensitivity
NGS of long-range PCR-amplified mtDNA	Very high sensitivity, can detect any point mutation or deletion	HQ, could miss large insertions
Whole exome sequencing	Can detect many mtDNA point mutations in addition to nuclear defects	HQ, incomplete coverage of mtDNA, potential interference by nuclear mtDNA sequences, semi-quantitative
Whole genome sequencing	Can potentially detect mtDNA point mutations, deletions and depletion in addition to nuclear defects	HQ, potential interference by nuclear mtDNA sequences, costly

Abbreviations: LQ, typically suitable for lesser-quality DNA (e.g., as in urine); HQ typically requires high-quality DNA

although most such patients also have mtDNA deletions. Testing by Southern blotting has largely been replaced by a combination of long-range PCR and quantitative PCR [37], which together provide a sensitive and quantitative detection method. While single deletions are usually sporadic [38], multiple mtDNA deletions are secondary to mutations in nuclear genes associated with mtDNA replication or nucleotide metabolism [19]. Mutations in this group of nuclear genes can also result in mtDNA depletion, which can be tested for by Southern blotting or quantitative PCR [39]. Single and multiple mtDNA deletions and mtDNA depletion typically show high tissue specificity and are often absent or at very low levels in blood of affected individuals [38, 40].

NGS methods can potentially detect mtDNA point mutations and deletions in a single assay, simplifying mutation detection. For mtDNA analysis, NGS can be performed on multiple overlapping PCR fragments or on a single long-range PCR amplicon covering the entire mtDNA coding sequence [32, 33]. Both strategies typically use a high depth of coverage so that each mtDNA is read at a depth over 1,000-fold, providing high sensitivity of detecting and quantifying even very small mutant loads. An advantage of a single PCR amplicon is that it provides even coverage of mtDNA across the genome, allowing sensitive detection of mtDNA deletions as well as point mutations. Whole exome sequencing involves targeted capture of DNA regions using RNA baits, and usually targets only nuclear coding regions. However, the high copy number of mtDNA means that mtDNA sequence variants can often be detected in whole exome data, if they are looked for [41, 42]. The coverage of mtDNA tends to be relatively low and nonuniform, so exome sequencing is typically not a robust method for detecting mtDNA point mutations or deletions. In contrast, whole genome sequencing contains high coverage of DNA sequence with robust detection of mtDNA point mutations and potentially of mtDNA deletions if appropriate algorithms are used. It is therefore possible that within coming years whole genome sequencing may prove an extremely useful method able to identify not only nuclear DNA mutations causing mitochondrial diseases but also mtDNA point mutations, deletions and depletions.

A detailed understanding of mtDNA heteroplasmy is essential for interpretation of diagnostic mtDNA analysis, in relation both to the specific mutation detected and to the tissue from which DNA was obtained. As a general rule, most pathogenic mtDNA mutations cause disease only when present at a high mutant load in clinically involved tissues [40, 43]. This pathogenic threshold may be as low as 30–50 percent mutant load for mtDNA deletions and some point mutations, but for others may be over 90 percent [44]. Indeed, for some mutations such as m.11778 G>A and m.1555A>G, most patients are homoplasmic for the mutation with no detectable wild-type mtDNA [43]. As mentioned earlier, the heteroplasmic mutant load of some mtDNA point mutations and deletions decreases in blood with increasing age. Pathogenic variants such as m.3243A>G and mtDNA deletions can therefore be detected at very low levels or may be undetectable in blood or buccal samples of affected adults, even though there may be a high level of the same mutation in muscle, brain or other tissues [35, 40]. In addition, at least 1/500 individuals in the population carry a pathogenic mtDNA variant, despite only 1/10,000 being diagnosed with mtDNA disease [45]. Large diagnostic centers will thus identify individuals carrying a pathogenic mtDNA mutation whose symptoms are not necessarily caused by the mutation. Additional caution needs to be applied when low levels of heteroplasmy are detected by exome or genome sequencing. The nuclear genome contains thousands of nuclear mtDNA sequences, often called NuMTs, with high homology to mtDNA. In capture-based NGS tests, such variants can be present at low levels of apparent heteroplasmy, and can potentially be confused with known or novel mtDNA mutations [41]. In summary, absence or low levels of certain mtDNA mutations in blood does not exclude causation and neither does their presence prove causation. Interpretation of common mtDNA mutations is usually straightforward, based on published experience. For novel mutations or variants of uncertain significance, confirmatory testing would typically involve measuring the level of heteroplasmy in urine or blood from family members to correlate mutant load with symptoms, together with studies to determine whether the biochemical consequences of the mutation are consistent with expectation. Contrary to expectations that pathogenic mtDNA point mutations typically show maternal inheritance, recent evidence suggests that *de novo* mtDNA point mutations are a common cause of mtDNA disease [46].

Nuclear DNA Mutations

Nuclear DNA analysis is typically performed by one of four approaches. Sanger sequencing of specific genes is usually appropriate only when the clinical and laboratory features suggest a specific syndrome such as Alpers syndrome, sensory ataxia neuropathy dysarthria and ophthalmoplegia (SANDO) or mitochondrial recessive ataxia syndrome (MIRAS) (all commonly caused by recessively inherited *POLG* mutations), mitochondrial neurogastrointestinal encephalomyopathy (MNGIE, *TYMP*), Barth syndrome (*TAZ*), Sengers syndrome (*AGK*), infantile-onset spinocerebellar ataxia (IOSCA, *C10orf2*) or leukoencephalopathy, brainstem and spinal cord involvement with lactate elevation (LBSL, *DARS2*) [30]. The disadvantage of this approach is that phenotypic overlap means that some patients suspected of such conditions may actually have mutations in another gene. Sanger sequencing typically ensures complete coverage of all coding exons, which is not always the case with NGS methods. However, with the possible exception of pathognomonic presentations such as those listed earlier, NGS methods will typically have a higher diagnostic yield than Sanger sequencing of single genes [47].

NGS gene panels provide a second approach to investigating nuclear genes and typically comprise tens to hundreds of known disease-causing genes [48]. These panels are typically cheaper and faster to analyze than whole exome or genome sequencing, can be run at high levels of sequence coverage (e.g., >200x) and can be optimized with additional capture probes to ensure high coverage of each exon [31, 49]. The main disadvantage of such panels is that tens of new mitochondrial disease genes are still being reported each year [19], so fixed panels require regular updating and revalidation in a diagnostic setting.

The third common approach that is becoming routine in many genomic diagnostic centers is whole exome sequencing, typically performed at ~100x sequence coverage [31, 33, 50]. Interpretation is often based on analysis of a virtual panel of a subset of genes to simplify filtering of true pathogenic variants and to avoid the potential for incidental findings not relevant to the condition being investigated. The diagnostic utility of whole exome sequencing continues to be improved with the development of "augmented" whole exome capture, using extra baits to ensure high coverage of 98 percent or more of exons in genes previously associated with disease [51]. A major advantage of the virtual panel approach is that for patients where no diagnosis is obtained with the initial gene panel, the computational analysis can be expanded to all known disease genes or to research analysis of the whole exome. Additionally, data can be re-interrogated bioinformatically as new disease genes are identified, rather than having to run an updated capture-based panel.

Whole genome sequencing is the fourth approach to diagnosis of mitochondrial disease, and both DNA heteroplasmy levels and copy number can be estimated from whole genome sequence data [52]. This approach offers the potential advantage of providing robust diagnosis of both nuclear and mtDNA variants in one analysis if DNA from an appropriate tissue is studied. However, as yet there do not appear to be any reports validating whole genome sequencing for diagnosis of mtDNA disease in a pathology context. For nuclear genes, avoiding library-based capture means that whole genome sequencing at 30x coverage should have more even coverage than a 100x whole exome, potentially detecting single nucleotide variants that may be missed by whole exomes, although a recent analysis suggested that augmented whole exomes have superior detection of variants in coding regions of known disease genes [51]. Whole genome sequencing is better able to detect copy number variants in nuclear genes than exome sequencing and can potentially identify sequence variants in intronic or "non-coding" regions, although these remain difficult to interpret [51].

Genomic diagnosis based on panels, exomes or genomes faces a multiple testing issue that traditional Sanger sequencing did not. If likely pathogenic variants are found in single gene candidates selected by the clinical or biochemical diagnosis, the likelihood of the variants being truly causative is relatively high. When genomic analyses test hundreds or thousands of genes, one expects to find numerous variants that are plausible candidates for causation, potentially in multiple genes. Extra caution is thus needed in defining the certainty of putative pathogenic variants identified by genomic analyses. A number of studies have suggested approaches to defining causality of variants, which typically relies on categorizing the type of mutation, algorithmic predictions of causation, population frequency of the variants, segregation of the variants within the family, previous publications and databases of pathogenic variants [31, 48, 53]. The American College of Medical Genetics has published standards and guidelines [54], which are useful to support decision-making as are resources developed by the MseqDR consortium [55]. Unless the variants show clear evidence of pathogenicity according to such criteria, it is highly desirable to discuss the evidence for causation in a multidisciplinary team meeting.

As genomic diagnosis increasingly becomes a first-line diagnostic test, it is particularly important to consider results in the context of what is known about the genomic architecture of the specific disease or enzyme defect. While some mitochondrial syndromes are caused by mutations in a relatively small number of genes, others are more diverse; for example, Leigh syndrome comprises more than 75 different monogenic disorders [56]. The relevance of previous studies can be illustrated by considering a case where novel mutations are found in a gene encoding one of the 44 complex I subunits:

- Pathogenic subunit mutations typically result in an enzyme defect in skin fibroblasts, so enzyme analysis of fibroblasts could provide further evidence of causation [57, 58];
- Patients with mutations in subunits such as *NDUFS4* or *NDUFS6* typically have two loss of function alleles and these mutations result in a "crippled" complex I missing the matrix arm when analyzed by Blue-Native PAGE [59];
- Knockout of the genes encoding the *NDUFV3, NDUFA7* and *NDUFA12* subunits results in little if any loss of complex I activity and no patients have been reported with mutations in these genes, suggesting even complete loss of function may not result in a complex I enzyme defect [59].

Thus, considering the type of mutations identified in a specific complex I subunit gene can prompt follow-up studies to establish causation and may cast doubt on the likelihood of the variants being truly pathogenic. Similar observations can be used to refine the likelihood of causation and follow-up studies for mutations in genes affecting other mitochondrial processes such as mtDNA maintenance and mtDNA translation, or disorders of other individual respiratory chain complexes (e.g., isolated complex IV deficiency) [31, 33, 48].

For those patients who remain unsolved after NGS analyses, it is important to consider potential technical limitations (e.g., exons with poor coverage or failure to detect copy number variants or small insertion/deletions) and bioinformatic assumptions made during filtering of the data. Mitochondrial disorders show considerable phenotypic overlap with other metabolic and neurogenetic disorders, and it is not uncommon to find diagnoses in genes not regarded as causing classical mitochondrial disorders [60]. It is also important to be careful about assumptions of the likely mode of inheritance. For example, most children with mitochondrial diseases have autosomal recessive disorders, and this is often assumed to be the case in consanguineous families, but mtDNA mutations can be causative in such families [61]. Similarly, there are a number of X-linked pediatric disorders (e.g., mutations in the *TAZ* and *AIFM1* genes) and other genes (e.g., *POLG* and *SLC25A4*) that have dominant mutations causing adult-onset CPEO and recessive mutations causing childhood-onset disease. A recent study has identified *de novo* dominant mutations in *SLC25A4* as a cause of severe neonatal myopathy and multiple respiratory chain defects in multiple families. This mechanism would not typically be thought of in early-onset mitochondrial disease [62], and as such challenges the way in which candidate variants from whole exome datasets should be prioritized informatically.

In conclusion, genomic testing is now able to identify the pathogenic basis of more than 50 percent of patients with mitochondrial disease, but should always be considered in the context of the available clinical features and laboratory investigations together with careful reassessment of cases where initial analyses fail to identify a likely molecular diagnosis. Diagnostic guidelines and criteria [2, 63] remain useful to consider the genomic evidence in the context of these wider features and to plan whether further investigations are necessary for validation.

References

1. McFarland, R., Taylor, R.W. and Turnbull, D. M. A neurological perspective on mitochondrial disease. *The Lancet Neurology* 2010; **9**: 829–840.

2. Parikh, S., Goldstein, A., Koenig, M. K., et al. Diagnosis and management of mitochondrial disease: A consensus statement from the Mitochondrial Medicine Society. *Genet Med* 2015; **17**: 689–701.

3. Johnson, M. A., Turnbull, D. M., Dick, D. J., et al. A partial deficiency of cytochrome c oxidase in chronic progressive external ophthalmoplegia. *J Neurol Sci* 1983; **60**: 31–53.

4. Taylor, R.W., Giordano, C., Davidson, M. M., et al. A homoplasmic mitochondrial transfer ribonucleic acid mutation as a cause of maternally inherited hypertrophic cardiomyopathy. *Journal of the American College of Cardiology* 2003; **41**: 1786–1796.

5. Zhu, Z., Yao, J., Johns, T., et al. SURF1, encoding a factor involved in the biogenesis of cytochrome c oxidase, is mutated in Leigh syndrome. *Nat Genet* 1998; **20**: 337–343.

6. Tiranti, V., Hoertnagel, K., Carrozzo, R., et al. Mutations of SURF-1 in Leigh disease associated with cytochrome c oxidase deficiency. *American Journal of Human Genetics* 1998; **63**: 1609–1621.

7. Papadopoulou, L. C., Sue, C. M., Davidson, M. M., et al. Fatal infantile cardioencephalomyopathy with COX deficiency and mutations in SCO2, a COX assembly gene. *Nat Genet* 1999; **23**: 333–337.

8. Bonilla, E., Sciacco, M., Tanji, K., et al. New morphological approaches to the study of mitochondrial encephalomyopathies. *Brain Pathology* (Zurich, Switzerland) 1992; **2**: 113–119.

9. Sciacco, M., Bonilla, E., Schon, E. A., et al. Distribution of wild-type and common deletion forms of mtDNA in normal and respiration-deficient muscle fibers from patients with mitochondrial myopathy. *Hum Mol Genet* 1994; **3**: 13–19.

10. Durham, S. E., Samuels, D. C., Cree, L. M., et al. Normal levels of wild-type mitochondrial DNA maintain cytochrome c oxidase activity for two pathogenic mitochondrial DNA mutations but not for m.3243A->G. *American Journal of Human Genetics* 2007; **81**: 189–195.

11. Muller-Hocker, J. Cytochrome-c-oxidase deficient cardiomyocytes in the human heart – an age related phenomenon. A histochemical ultracytochemical study. *The American Journal of Pathology* 1989; **134**: 1167–1173.

12. Bender, A., Krishnan, K. J., Morris, C. M., et al. High levels of mitochondrial DNA deletions in substantia nigra neurons in aging and Parkinson disease. *Nat Genet* 2006; **38**: 515–517.

13. Rygiel, K. A., Tuppen, H. A., Grady, J. P., et al. Complex mitochondrial DNA rearrangements in individual cells from patients with sporadic inclusion body myositis. *Nucleic Acids Res* 2016; **44**: 5313, 5329.

14. Taylor, R. W., Barron, M. J., Borthwick, G. M., et al. Mitochondrial DNA mutations in human colonic crypt stem cells. *The Journal of Clinical Investigation* 2003; **112**: 1351–1360.

15. Tanji, K. and Bonilla, E. Optical imaging techniques (histochemical, immunohistochemical, and in situ hybridization staining methods) to visualize mitochondria. *Methods in Cell Biology* 2007; **80**: 135, 154.

16. Rahman, S., Lake, B. D., Taanman, J. W., et al. Cytochrome oxidase immunohistochemistry: Clues for genetic mechanisms. *Brain* 2000; **123**: Pt. 3, 591–600.

17. Rocha, M. C., Grady, J. P., Grunewald, A., et al. A novel immunofluorescent assay to investigate oxidative phosphorylation deficiency in mitochondrial myopathy: Understanding mechanisms and improving diagnosis. *Scientific Reports* 2015; **5**: 15037.

18. Luft, R., Ikkos, D., Palmieri, G., et al. A case of severe hypermetabolism of non-thyroid origin with a defect in the maintenance of mitochondrial respiratory control. A correlated clinical, biochemical and morphological study. *J Clin Invest* 1962; **41**: 1776–1804.

19. Mayr, J. A., Haack, T. B., Freisinger, P., et al. Spectrum of combined respiratory chain defects. *Journal of Inherited Metabolic Disease* 2015; **38**: 629–640.

20. Kirby, D. M., Thorburn, D. R., Turnbull, D. M., et al. Biochemical assays of respiratory chain complex activity. *Methods in Cell Biology* 2007; **80**: 93–119.

21. Rodenburg, R. J., Schoonderwoerd, G. C., Tiranti, V., et al. A multi-center comparison of diagnostic methods for the biochemical evaluation of suspected mitochondrial disorders. *Mitochondrion* 2013; **13**: 36–43.

22. Medja, F., Allouche, S., Frachon, P., et al. Development and implementation of standardized respiratory chain spectrophotometric assays for clinical diagnosis. *Mitochondrion* 2009; **9**: 331–339.

23. Invernizzi, F., D'Amato, I., Jensen, P. B., et al. Microscale oxygraphy reveals OXPHOS impairment in MRC mutant cells. *Mitochondrion* 2012; **12**: 328–335.

24. Bonnen, P. E., Yarham, J. W., Besse, A., et al. Mutations in FBXL4 cause mitochondrial encephalopathy and a disorder of mitochondrial DNA maintenance. *American Journal of Human Genetics* 2013; **93**: 471–481.

25. McKenzie, M., Lazarou, M., Thorburn, D. R., et al. Analysis of mitochondrial subunit assembly into respiratory chain complexes using Blue Native polyacrylamide gel electrophoresis. *Anal Biochem* 2007; **364**: 128–137.

26. Lim, S. C., Smith, K. R., Stroud, D. A., et al. A founder mutation in *PET100* causes isolated complex IV deficiency in Lebanese individuals with Leigh syndrome. *American Journal of Human Genetics* 2014; **94**: 209–222.

27. Thompson Legault, J., Strittmatter, L., Tardif, J., et al. A metabolic signature of mitochondrial dysfunction revealed through a monogenic form of Leigh syndrome. *Cell Rep* 2015; **13**: 981–989.

28. Floyd, B. J., Wilkerson, E. M., Veling, M. T., et al. Mitochondrial protein interaction mapping identifies regulators of respiratory chain function. *Mol Cell* 2016; **63**: 621–632.

29. Frazier, A. E., Thorburn, D. R. and Compton, A. G. Mitochondrial energy generation disorders: genes, mechanisms and clues to pathology. *Journal of Biological Chemistry* 2018 doi: 10.1074/jbc. R117.809194. [Epub ahead of print].

30. Kirby, D. M. and Thorburn, D. R. Approaches to finding the molecular basis of mitochondrial oxidative phosphorylation disorders. *Twin Res Hum Genet* 2008; **11**: 395–411.

31. Carroll, C. J., Brilhante, V. and Suomalainen, A. Next-generation sequencing for mitochondrial disorders. *British Journal of Pharmacology* 2014; **171**: 1837–1853.

32. Zhang, W., Cui, H. and Wong, L. J. Comprehensive one-step molecular analyses of mitochondrial genome by massively parallel sequencing. *Clinical Chemistry* 2012; **58**: 1322–1331.

33. Kohda, M., Tokuzawa, Y., Kishita, Y., et al. A comprehensive genomic analysis reveals the genetic landscape of mitochondrial

respiratory chain complex deficiencies. *PLoS genetics* 2016; **12**: e1005679.

34. Rahman, S., Poulton, J., Marchington, D., et al. Decrease of 3243 A->G mtDNA mutation from blood in MELAS syndrome: A longitudinal study. *American Journal of Human Genetics* 2001; **68**: 238–240.

35. Whittaker, R. G., Blackwood, J. K., Alston, C. L., et al. Urine heteroplasmy is the best predictor of clinical outcome in the m.3243A>G mtDNA mutation. *Neurology* 2009; **72**: 568–569.

36. Broomfield, A., Sweeney, M. G., Woodward, C. E., et al. Paediatric single mitochondrial DNA deletion disorders: An overlapping spectrum of disease. *Journal of Inherited Metabolic Disease* 2015; **238**: 445–457.

37. He, L., Chinnery, P. F., Durham, S. E., et al. Detection and quantification of mitochondrial DNA deletions in individual cells by real-time PCR. *Nucleic Acids Res* 2002; **30**: e68.

38. Chinnery, P. F., DiMauro, S., Shanske, S., et al. Risk of developing a mitochondrial DNA deletion disorder. *Lancet* 2004; **364**: 592–596.

39. Dimmock, D., Tang, L. Y., Schmitt, E. S., et al. Quantitative evaluation of the mitochondrial DNA depletion syndrome. *Clinical Chemistry* 2010; **56**: 1119–1127.

40. DiMauro, S. Mitochondrial encephalomyopathies – fifty years on: The Robert Wartenberg Lecture. *Neurology* 2013; **81**: 281–291.

41. Ye, F., Samuels, D. C., Clark, T., et al. High-throughput sequencing in mitochondrial DNA research. *Mitochondrion* 2014; **17**: 157–163.

42. Griffin, H. R., Pyle, A., Blakely, E. L., et al. Accurate mitochondrial DNA sequencing using off target reads provides a single test to identify pathogenic point mutations. *Genet Med* 2014; **16**: 962, 971.

43. Saneto, R. P. and Sedensky, M. M. Mitochondrial disease in childhood: mtDNA encoded. *Neurotherapeutics* 2013; **10**: 199–211.

44. Kirby, D. M., Boneh, A., Chow, C. W., et al. Low mutant load of mitochondrial DNA G13513A mutation can cause Leigh disease. *Ann Neurol* 2003; **54**: 473–478.

45. Gorman, G. S., Schaefer, A. M., Ng, Y., et al. Prevalence of nuclear and mtDNA mutations related to adult mitochondrial disease. *Annals of Neurology*. 2015; May; **77** (5):753–759.

46. Sallevelt, S. C., de Die-Smulders, C. E., Hendrickx, A. T., et al. De novo mtDNA point mutations are common and have a low recurrence risk. *Journal of Medical Genetics*. 2017 Feb; **54**(2):73–83.

47. Neveling, K., Feenstra, I., Gilissen, C., et al. A post-hoc comparison of the utility of sanger sequencing and exome sequencing for the diagnosis of heterogeneous diseases. *Hum Mutat* 2013; **34**: 1721–1726.

48. Calvo, S. E., Tucker, E. J., Compton, A. G., et al. High-throughput, pooled sequencing identifies mutations in NUBPL and FOXRED1 in human complex I deficiency. *Nat Genet* 2010; **42**: 851–858.

49. Platt, J., Cox, R. and Enns, G. M. Points to consider in the clinical use of NGS panels for mitochondrial disease: An analysis of gene inclusion and consent forms. *Journal of Genetic Counseling* 2014; **23**: 594–603.

50. Haack, T. B., Danhauser, K., Haberberger, et al. Exome sequencing identifies ACAD9 mutations as a cause of complex I deficiency. *Nat Genet* 2010; **42**: 1131–1134.

51. Ashley, E. A. Towards precision medicine. *Nat Rev Genet* 2016; **17**: 507–522.

52. Ding, J., Sidore, C., Butler, T. J., et al. Assessing mitochondrial DNA variation and copy number in lymphocytes of ~2,000 Sardinians using tailored sequencing analysis tools. *PLoS genetics* 2015; **11**, e1005306.

53. MacArthur, D. G., Manolio, T. A., Dimmock, D. P., et al. Guidelines for investigating causality of sequence variants in human disease. *Nature* 2014; **508**: 469–476.

54. Richards, S., Aziz, N., Bale, S., et al. Standards and guidelines for the interpretation of sequence variants: A joint consensus recommendation of the

American College of Medical Genetics and Genomics and the Association for Molecular Pathology. *Genet Med* 2015; **17**: 405–423.

55. Falk, M. J., Shen, L., Gonzalez, M. et al. Mitochondrial Disease Sequence Data Resource (MSeqDR): A global grass-roots consortium to facilitate deposition, curation, annotation, and integrated analysis of genomic data for the mitochondrial disease clinical and research communities. *Molecular Genetics and Metabolism* 2015; **114**: 388–396.

56. Lake, N. J., Compton, A. G., Rahman, S., et al. Leigh syndrome: One disorder, more than 75 monogenic causes. *Annals of Neurology* 2016; **79**: 190–203.

57. Hoefs, S. J., Rodenburg, R. J., Smeitink, J. A., et al. Molecular base of biochemical complex I deficiency. *Mitochondrion* 2012; **12**: 520–532.

58. Fassone, E. and Rahman, S. Complex I deficiency: Clinical features, biochemistry and molecular genetics. *Journal of Medical Genetics* 2012; **49**: 578–590.

59. Stroud, D. A., Surgenor, E. E., Formosa, L. E., et al. Accessory subunits are integral for assembly and function of human mitochondrial complex I. *Nature* 2016; Oct 6; **538**(7623): 123–126.

60. Lieber, D. S., Calvo, S. E., Shanahan, K., et al. Targeted exome sequencing of suspected mitochondrial disorders. *Neurology* 2013; **80**: 1762–1770.

61. Alston, C. L., He, L., Morris, A. A., et al. Maternally inherited mitochondrial DNA disease in consanguineous families. *Eur J Hum Genet* 2011; **19**: 1226–1229.

62. Thompson, K., Majd, H., Dallabona, C., et al. Recurrent de novo dominant mutations in SLC25A4 cause severe early-onset mitochondrial disease and loss of mitochondrial DNA copy number. *American Journal of Human Genetics* 2016; Dec 1; **99**(6): 1405.

63. Bernier, F. P., Boneh, A., Dennett, X., et al. Diagnostic criteria for respiratory chain disorders in adults and children. *Neurology* 2002; **59**: 1406–1411.

Chapter 5

Clinical Genetics of Mitochondrial Diseases

Marni J. Falk and Colleen Clarke Muraresku

Clinical Genetics of Mitochondrial Diseases

Genetic Counseling: Genetic counseling is challenging in mitochondrial disease due to the wide number of genetic etiologies, variability in diagnostic testing algorithms, extensive clinical variability and limited preventative, management and treatment options for patients and their families [1]. Primary mitochondrial disease implies a genetic etiology exists or is suspected, with any mode of inheritance possible (Figure 5.1). Indeed, inheritance patterns for mitochondrial disease can be maternal (meaning due to a pathogenic mutation in mitochondrial DNA [mtDNA]), autosomal recessive, autosomal dominant, X-linked or sporadic (*de novo*). Each mode of inheritance bears a different recurrence risk for siblings (full or half) and offspring of an affected individual. The majority of mitochondrial diseases having symptom onset in childhood (67 percent to 90 percent) is inherited in an autosomal recessive manner [2]. Recurrence risk for most autosomal recessive disorders is low should either asymptomatic carrier parent have a different partner for future pregnancies, assuming a low background frequency in the population of the causative gene disorder. In maternally inherited (mtDNA) or X-linked mitochondrial diseases, a carrier or affected mother's recurrence risk is independent of her partner. Similarly, in autosomal dominant disorder, affected individuals have a one in two (50 percent) likelihood of passing on the disorder to each child regardless of their partner. When the precise genetic etiology for the affected individual is unknown and no clear inheritance patterns emerges by pedigree analysis, definitive recurrence risk cannot be provided for the specific mitochondrial disease occurring in a given family. Thus, it is important to establish a precise molecular diagnosis to accurately inform the recurrence risk counseling for the parents of an affected child or for the affected individual themselves.

Family History: Collecting a detailed family history of medical symptoms and signs is essential when taking a medical history. Indeed, the medical pedigree often serves as a useful diagnostic and risk assessment tool. Given the variability of symptoms in mitochondrial disease, obtaining a pedigree that begins with the proband (affected individual) and obtaining information about their first-, second- and even third-degree relatives can allow the provider to connect symptoms that may not otherwise appear related to one another [11] (see Chapters 2 and 3 for further information of clinical history taking in susepected cases of mitochondrial disease). This information often serves as a guide for determining which diagnostic testing to pursue in the proband to further evaluate for a mitochondrial disease. For example, obtaining a maternal family history of multiple individuals with features of mitochondrial disease would suggest the importance of first

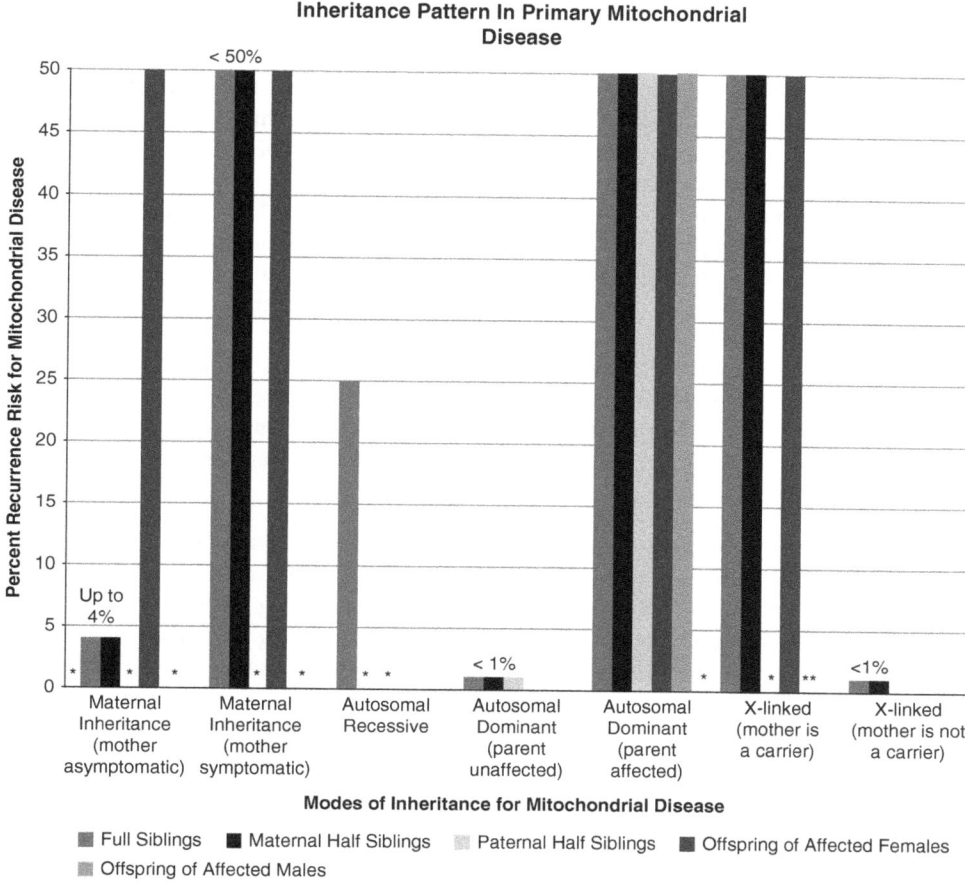

Figure 5.1 * indicates no recurrence risk. (A black and white version of this figure will appear in some formats. For the color version, please refer to the plate section.)

pursuing mtDNA sequencing and deletion/duplication studies before pursuit of nDNA-based sequencing studies. The absence of a clear maternal family history of mitochondrial disease symptoms, or the identification of a likely Mendelian pattern of disease inheritance, would prioritize nDNA testing as the most appropriate course, although it remains possible that mtDNA mutations can arise *de novo* or at higher heteroplasmy levels in an affected individual than may be present in asymptomatic family members [1].

mtDNA-based Mitochondrial Disease: While males and females both have mtDNA, an individual's mtDNA is exclusively inherited from one's mother. Both oocytes and sperm have mtDNA, but the mtDNA from sperm are actively excluded from the fertilized oocyte. mtDNA diseases can result from mtDNA point mutations, small deletions or large deletions that may or may not encompass duplications. To properly evaluate and counsel a family who has been diagnosed with or is suspected to have a mitochondrial disease, it is important to correctly interpret what testing was completed and to determine whether the patient's mother is symptomatic. Accurate diagnostic testing for mtDNA mutations may require tissue-specific testing, such as muscle or urine, rather than blood. When a child with a negative maternal family history is diagnosed with an mtDNA-based mitochondrial disease, it is likely the mutation or deletion occurred *de novo* (new) in the affected child [3]. This possibility would yield a low recurrence risk for the family, on the order of 1 percent to 4 percent for all full siblings and maternal half-siblings based on the possibility of gonadal mosaicism (meaning more than one oocyte might harbor the mutation). When a mother is herself symptomatic, the empiric risk is up to a one in two (50 percent) chance that each of her future offspring will also be affected [3]. Disease examples caused by point mutations include mitochondrial encephalomyopathy with lactic acidosis and stroke-like episodes (MELAS), and myoclonic epilepsy with ragged red fibers (MERRF), although more than 300 confirmed pathogenic mutations causing a wide variety of clinical manifestations have now been identified. Deletions in the mtDNA cause a variety of clinical manifestations depending on the tissues in which they arise, including Pearson syndrome, chronic progressive external ophthalmoplegia (CPEO) and Kearns-Sayre syndrome (KSS).

nDNA-based Mitochondrial Disease: The majority of mitochondrial disease is caused by nDNA mutations. Mutations in more than 250 nuclear genes have been shown to cause mitochondrial disease [4], with many new causative nuclear genes still being recognized on a regular basis with the increasing utilization of next-generation sequencing methodologies. No one gene accounts for more than a few percent of mitochondrial diseases, with the majority of disorders being individually rare and caused by a variety of potential mutations within a given gene. When pathogenic nDNA mutation(s) are identified in a family, it is important to determine whether the mode of inheritance is autosomal recessive, autosomal dominant or X-linked. In any scenario, the maximal recurrence risk does not exceed one in two (50 percent) (see Figure 5.1). As the majority of childhood-onset mitochondrial disease is inherited in an autosomal recessive manner, the recurrence risk for the parents to have another affected child with an autosomal recessive disease is one in four (25 percent) with each pregnancy [2]. Siblings of the affected child who are not affected should be counseled that they are each at a two in three (67 percent) risk of being asymptomatic carriers for the condition. Genetic testing to confirm carrier status is not offered to minors; adult relatives can undergo genetic testing to determine their carrier status. Unrelated partners of known carriers should be offered full sequencing analysis of the disease gene, ideally prior to pregnancy, to determine if they carry any disease-causing mutation that would put the couple at a one in four (25 percent) risk for each of their offspring to be affected. Examples of nDNA gene disorders that are inherited in an autosomal recessive manner include *POLG, TK2, DGUOK* and *SURF1*.

Mitochondrial disease can also be caused by nDNA gene mutations that occur in an autosomal dominant fashion. When no family history of disease symptoms exists, the mutation likely occurred *de novo* (new) in the affected individual. This possibility yields a low recurrence risk of less than 1 percent due to gonadal mosaicism. For an affected individual, each of their future children is at a one in two (50 percent) risk of inheriting the disease gene mutation and also being affected. Reduced and/or variable penetrance can occur in many autosomal dominant disorders. Disease examples caused by nDNA gene mutations that may be inherited in an autosomal dominant manner include progressive external ophthalmoplegia (PEO), which may result from mutations in a range of genes such as *POLG* or *C10orf2*.

A final mode of Mendelian inheritance that may cause mitochondrial disease is X-linked inheritance. In X-linked conditions, males are typically more severely affected than females. This relates largely to males having only one X chromosome, thereby manifesting any genetic disorders it may carry. In contrast, females who may be carriers of a mutation may be either clinically unaffected or present with manifestations of the disease. This variation in clinical presentation relates to the phenomenon of X-inactivation, which is a random process in each female cell that turns off expression of one X chromosome. When a female is a carrier for an X-linked genetic condition, 50 percent of her sons will be affected and 50 percent of her daughters will be carriers. If a male affected with an X-linked disorder has children, all of his daughters will be carriers while none of his sons (who will inherit only his Y chromosome) will develop the disease [2]. Mitochondrial disease examples inherited in an X-linked fashion are Barth syndrome that is caused by *TAZ* mutations and pyruvate dehydrogenase deficiency that results from *PDHA1* mutations.

Prognostic Counseling: Once a specific genetic diagnosis is confirmed, counseling should be provided about the specific range of clinical variability known to occur in other individuals with that condition. Within a family affected by a known pathogenic mutation in either an mtDNA or nDNA gene, prognosis is dependent on several factors. For mtDNA gene mutations, both heteroplasmy load and threshold effect within a given tissue must be considered to predict clinical outcomes. As a mtDNA mutation may be present in only a percentage of one's cells or tissues, "heteroplasmy load" is used to describe the ratio of the mutated mtDNA genomes relative to normal mtDNA genomes in any one tissue (blood, muscle, liver, urine) [5]. Heteroplasmy load can often vary between different tissue types in the same individual, making prognostic predictions more challenging since it is not possible to quantify the mutation load in less clinically accessible tissues (brain, heart, etc.). The threshold effect refers to the minimal percentage of normal mtDNA that is needed to support the healthy function of a given organ. For example, one sibling born with a high mtDNA mutation load may die of their disease from complications of metabolic stroke or epilepsy in early childhood, while another sibling within the same family who has a lower mutation load may not develop symptoms until later in life and present with more variable symptoms such as migraines, sensorineural hearing loss, vision loss, or diabetes mellitus.

In proven nDNA-based mitochondrial disorders, clinical manifestations can similarly be quite variable even though the genetic mutation is present in every cell of the body. However, some disorders are more severe in all affected cases, such as those that cause Leigh syndrome involving deep brain strokes, hypotonia and epilepsy. For autosomal dominant or X-linked disorders, children of affected individuals may often present across a wide spectrum of clinical manifestations and age of onset that are difficult to predict. Modifier genes across both nuclear and mitochondrial genomes, as well as lifestyle factors, often play into the penetrance or severity of a given disorder.

Disease Prevention

Disease prevention is available for mitochondrial disease only if a known pathogenic mutation(s) is first identified either in an nDNA or mtDNA gene [6]. Prenatal genetic testing for nDNA gene mutations is more straightforward and established, as the clinical outcomes in mtDNA mutations depend on factors such as heteroplasmy and threshold effect that may widely vary over time [7]. Predictive genetic testing can be used to determine whether or not an embryo or a fetus carries a given mutation(s). However, preimplantation (testing the embryo) or prenatal (testing the fetus) genetic diagnostic techniques are not 100 percent reliable. Thus, confirmatory genetic diagnostic testing is still recommended once the child is born. Additional options that should be discussed with couples who harbor known pathogenic nDNA gene mutations and want to have an unaffected child include adoption as well as gamete (egg or sperm) donation using donors who do not carry the same genetic disorder.

Genetic testing for mtDNA mutations either prior to or during a pregnancy is complicated due to the maternal mode of inheritance and heteroplasmy [5]. When pursuing preimplantation genetic diagnosis at the time of in vitro fertilization (see below) for mtDNA mutation load determination in a three- or five-day embryo prior to implantation, it is difficult to extrapolate that the mutation load identified in the one or two embryonic cells tested will be reflective of the future heteroplasmic load in all of the different tissues of the resulting fetus or child. These same concerns also limit the utility of mtDNA mutation load heteroplasmy determination in the prenatal setting. For most women who are known to carry a pathogenic mtDNA mutation, they will not have many, if any at all, embryos with a zero mutation load. This would leave couples who undergo this procedure with the option to choose an embryo having the lowest mutation load relative to other embryos, but this does not guarantee a healthy outcome for the future child [8]. Selection of a male embryo may be considered as the resulting child, even if clinically affected, would not be faced with a similar recurrence risk for their own offspring, since mtDNA is passed on only through the maternal (oocyte) line. Newer methodologies that would enable prospective parents to reduce the mtDNA mutation load in affected oocytes or embryos to reliably below 1 or 2 percent are now being evaluated in the research setting [9]. However, at the current time the wide spectrum of clinical variability and variability that may occur over time in heteroplasmic load limit the predictive accuracy of preimplantation and prenatal genetic diagnostic options for mtDNA disorders.

If a precise molecular diagnosis has not yet been established in a family with a suspected mitochondrial disease, limited options exist to assure future biological children will not inherit the disease [1]. Family planning options that completely reduce risk remains adoption of nonbiological children. However, when no causative mutation or inheritance pattern is identified, having donor sperm or egg does not completely remove the recurrence risk in the family since it is not known whether the other parent may harbor the disease mutation. Even though the majority of pediatric mitochondrial disease is inherited in an autosomal recessive fashion, only confirmation of a precise molecular diagnosis can reliably exclude other possible modes of inheritance. Once a molecular diagnosis is confirmed in the affected individual, then the full range of preimplantation or prenatal genetic diagnostic testing options will exist to enable the family to have future children without mitochondrial disease [5].

In vitro fertilization (IVF) with preimplantation genetic diagnosis (PGD): IVF with PGD is a technique that requires the harvesting of gametes (sperm and oocytes) from both intended parents. The harvested sperm and eggs are fertilized in a petri dish, with typically multiple zygotes generated in each IVF cycle. The embryo then divides to reach a multicellular stage that varies between clinical centers from three-day embryos (approximately eight cells) to five-day embryos (approximately 100–150 cells), where embryos that make it to the five-day stage are likely to yield greater pregnancy rates [7]. At this point, one or two cells are removed from each embryo to test for the specific pathogenic mutation(s) that have been previously identified in the family. Once these results are available, only embryos found to not carry the disease mutation(s) are implanted in the mother's uterus. Additional embryos found to be unaffected can be frozen for implantation in future cycles. Affected embryos can be discarded or donated to research per the couple's request. Once a pregnancy is conceived, it is recommended that further prenatal genetic diagnostic testing be pursued during the pregnancy to have more accurate diagnostic testing. Such testing can either be done by invasive testing (chorionic villus sampling or amniocentesis, see below), or potentially with newer noninvasive maternal blood-based diagnostic testing options now being investigated for single gene disorders in the research setting [10].

IVF with PGD has risks associated with the procedure similar to those for any IVF process [7]. Along with these, other problems can arise from genetic testing in embryos:

1) There may not be any unaffected embryos identified. In other words, all embryos tested may carry the mutation(s).
2) Most individuals who pursue PGD do not have fertility issues, but still may not produce enough eggs to generate a chromosomally normal, mutation-free embryo to transfer.
3) Not enough eggs will fertilize to continue the process and warrant genetic testing.
4) Damage to the embryo may occur from removing one or more cells.
5) Genetic diagnostic testing is not 100 percent reliable.

Chorionic Villi Sampling (CVS): CVS is an invasive procedure performed at 10–12 weeks' gestation during the first trimester of pregnancy [11]. During this procedure, a few cells are removed that surround the fetus, which are known as chorionic villi cells and later develop into placenta. These cells can be removed by either trans-abdominal or trans-cervical approach and then cultured in a laboratory setting. They are then tested to determine if they carry the specific mutation(s) identified in the family. If the tested cells have the mutation, this is interpreted as the fetus also having the mutation, as the genetic makeup of the chorionic villi and the fetus are identical. However, there are cases where there is confined placental mosaicism, which means there is a mutation or genetic abnormality present in the placenta but not in the fetus. There have been no reported cases of confined placental mosaicism occurring in mitochondrial disease. Typically, test results are available to the family two to six weeks post-CVS sampling. The parent(s) can then use this information to guide their decision to continue or to terminate the pregnancy.

Amniocentesis: Amniocentesis is an invasive procedure performed between 16 and 20 weeks' gestation during the second trimester of pregnancy [11]. An amniocentesis uses a needle biopsy procedure that enters through the mother's abdomen to remove a small portion of the amniotic fluid that surrounds the fetus. Within amniotic fluid are cells that have sloughed off from the fetus, thus having the same genetic material as the fetus. These cells can be cultured or sent directly to a diagnostic testing laboratory to determine if they harbor the mutation(s) identified in the family. Testing results are typically reported to families within approximately two weeks post-sampling. The parent(s) can then use this information to guide their decision to continue or to terminate the pregnancy.

References

1. Vento J, Pappa B. Genetic counselling in mitochondrial disease. *Neurotherapeutics* 2013; **10**:243–250.

2. Falk M. Neurodevelopmental manifestations of mitochondrial disease. *J Dev Behav Pediatr* 2010; **31**:610–621.

3. Hellebrekers DMEI, Wolfe R, Hendrickx ATM, et al. PGD and heteroplasmic mitochondrial DNA point mutations: A systematic review estimating the chance of healthy offspring. *Hum Reprod Update* 2012; **18**(4):341–349.

4. Koopman WJH, Willems PHGM and Smeitink JAM. Monogenic mitochondrial disorders. *N Engl J Med* 2012; **366**:1132–1141.

5. Mitalipov S, Amato P, Parry S and Falk M. Limitations of preimplantation genetic diagnosis for mitochondrial DNA diseases. *Cell Reports* 2014; **7**:935–937.

6. Bredenoord AL, Pennings G, Smeets HJ, et al. Dealing with uncertainties: Ethics of prenatal diagnosis and preimplantation genetic diagnosis to prevent mitochondrial disorders. *Human Reproduction Update* 2008; **14**:83–94.

7. Bredenoord A, Dondorp W, Pennings G, et al. Preimplantation genetic diagnosis for mitochondrial DNA disorders: Ethical guidance for clinical practice. *European Journal of Human Genetics* 2009; **17**:1550–1559.

8. Sallevelt SCEH, Dreesen JCFM, Drüsedau M, et al. Preimplantation genetic diagnosis in mitochondrial DNA disorders: Challenge and success. *J Med Genet* 2013; **50** (1):125–132.

9. Buyx AM, Strech D and Schmidt H. Ethical issues raised by direct-to-consumer personal genome analysis and whole body scans: Discussion and contextualisation of a report by the Nuffield Council on Bioethics. *Z Evid Fortbild Qual Gesundhwes* 2012; **106**(1):29–39.

10. Liao GJ, Gronowski AM and Zhao Z. Non-invasive prenatal testing using cell-free fetal DNA in maternal circulation. *Clin Chim Acta* 2014; **428**:44–50.

11. Nesbitt V, Alston CL, Blakely EL, et al. A national perspective on prenatal testing for mitochondrial disease. *European Journal of Human Genetics* 2014; **22**:1255–1259.

Clinical Management
of Mitochondrial Diseases

Rita Horvath

Managing Complications – General Principles

Mitochondrial disorders are a heterogeneous group of diseases affecting different organs like brain, muscle, liver and heart. The severity of the disease is very variable and many patients require surveillance follow-up over their lifetime, often involving multiple disciplines. Although our understanding of the genetic defects and their pathological impact underlying mitochondrial diseases has significantly improved over the past decade, this has not been paralleled with regards to treatment. Currently, pharmacological treatment exists for only a very small number of patients with mitochondrial dysfunction due to rare metabolic defects. However, several new pharmacological agents with potential benefit for a wider group of mitochondrial diseases are currently being investigated *in vitro* (in human cells) and *in vivo* (in animal models) [1]. Nonpharmacological treatments including dietary interventions, organ transplantation, exercise and gene therapy are increasingly being studied. Management is primarily aimed at minimizing disability, preventing complications and providing prognostic information and genetic counseling based on current best practice.

Clinical follow-up: Because of the rarity and complexity of mitochondrial diseases, it is advisable that patients are referred to specialized centers with experience in mitochondrial disease. These centers usually offer regular (six monthly, yearly) follow-up examinations and guide local health professionals (neurologists, pediatricians, general practitioners, etc.) if the patients require treatment or interventions that can be dealt with locally.

The following examinations are recommended at regular follow-up of patients with various mitochondrial diseases (Figure 6.1): routine clinical examination, neurological examination (including vision, hearing, cognition), routine laboratory parameters including blood sugar, liver and kidney function. Because quite a few mitochondrial diseases can affect the heart, regular ECG with 24-hour monitoring and echocardiography is recommended. Regarding the most common diseases, the risk of cardiac arrhythmia is very high in Kearns-Sayre syndrome, and cardiomyopathy is frequent in MELAS. Repeated neuroimaging, electrophysiological tests, monitoring gastrointestinal symptoms or specific laboratory tests may be required in some patients depending on the diagnosis.

Genetic counseling with explanations of inheritance pattern, recurrence risk and prenatal/preimplantation genetic diagnostic testing is an important part of the regular consultations in patients with mitochondrial disease (see Chapter 5). Based on the improvement in genetic diagnosis of patients, and on recent natural history studies in mitochondrial diseases, the disease prognosis can be better estimated. The possibilities to prevent the transmission of

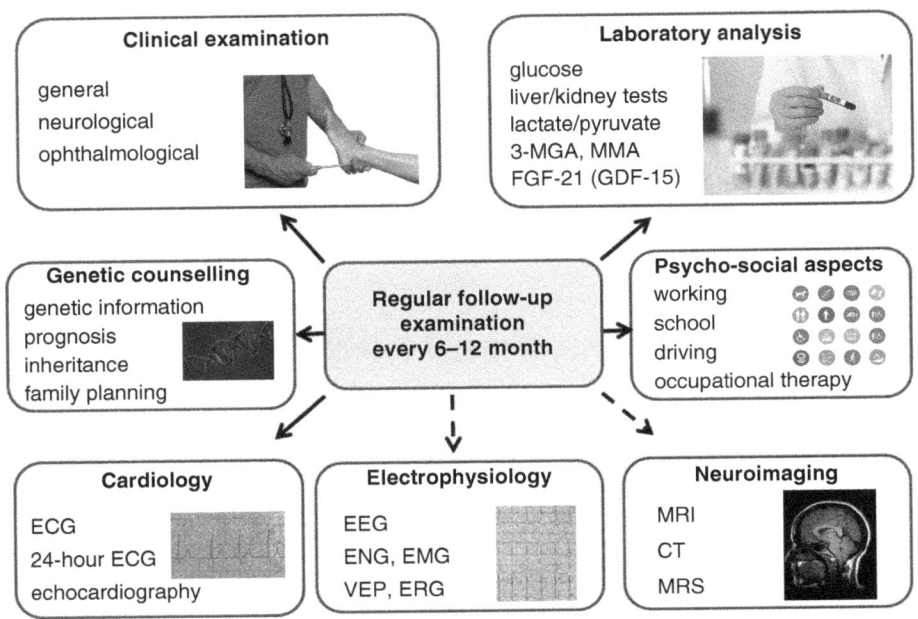

Figure 6.1 Regular follow-up of patients with mitochondrial disease. (A black and white version of this figure will appear in some formats. For the color version, please refer to the plate section.)

mitochondrial disease have been rapidly improving over the past few years; these options should be explained and discussed with the patients and families.

Psychological, social and environmental issues of patients with mitochondrial disease can also be better addressed in a specialized mitochondrial clinic.

Patients with mitochondrial disease should carry a surveillance card that details their underlying condition and provides management recommendations, and a Medic Alert bracelet is recommended.

Disease-Modifying Factors

Potentially Beneficial Therapies Currently Available in Mitochondrial Disease

Riboflavin metabolism defects: Riboflavin, or vitamin B_2, is the precursor of flavin adenine dinucleotide (FAD) and flavin mononucleotide (FMN), which are essential cofactors of numerous dehydrogenases involved in fatty acid beta-oxidation, branched chain amino acid catabolism, the mitochondrial electron transport chain and the tricarboxylic acid cycle [2]. Genetic defects of the riboflavin transport and riboflavin responsive mitochondrial diseases have been recently identified (Brown-Vialetto-Van Laere [BVVL] and Fazio-Londe syndromes, multiple acyl-CoA dehydrogenase deficiency [MADD], some *FLAD1* and *ACAD9* mutations). In patients with these conditions, a clinical and biochemical improvement has been shown due to oral riboflavin therapy. *SLC52A2* (less frequently SLC52A1 and SLC52A3) mutations cause severe, usually childhood-onset neuropathy, optic atrophy, deafness with frequent involvement of other cranial nerves due to reduced riboflavin uptake and reduced riboflavin transporter protein expression. A clinical and biochemical improvement has been shown on high-dose oral riboflavin therapy (10 mg/kg/day and sequentially increased to 50 mg/kg/day in pediatric patients and 1,500 mg/day in adult patients), particularly when initiated soon after the onset of symptoms. Mutations in the electron-transferring flavoprotein genes (*ETFA/ETFB*) and more frequently in its dehydrogenase (*ETFDH*) are causative for multiple acyl-CoA dehydrogenase deficiency. The majority of these patients show significant clinical improvement on riboflavin (100–400 mg/day), and an additional positive effect of ubiquinone (CoQ_{10}) has been reported [3]. *FLAD1* mutations were recently reported as a new genetic form of multiple acyl-CoA dehydrogenase deficiency [4]. Some patients with recessive *FLAD1* mutations present with severe and fatal disease in early childhood, but another group of patients with missense mutations has an adult-onset riboflavin responsive myopathy, similar to *ETFDH* mutations [4]. In addition, some patients carrying mutations in the *ACAD9* gene encoding a complex I assembly protein may benefit from supplementation with riboflavin [5)]. An increase of multiple acyl-carnitines in serum and urine may be the diagnostic biomarker of a potential riboflavin responsive condition.

Mutations in other mitochondrial transporters: Mutations in other mitochondrial transporter proteins causing potentially treatable mitochondrial disease have been recently identified. Biotin-responsive basal ganglia disease (BBGD) is an autosomal recessive inherited neuro-metabolic disorder characterized by subacute encephalopathy, dysarthria and dysphagia with occasional facial nerve palsy or ophthalmoplegia. Patients may progress to severe cogwheel rigidity, dystonia and quadriparesis [6]. Both biotin and thiamine are essential for disease management and supplementation results in clinical benefit, therefore clinicians should consider this disorder in any patients with subacute/acute encephalopathy, ataxia triggered by febrile illness and basal ganglia involvement [6].

Thiamine-responsive megaloblastic anemia syndrome (TRMA) is the association of diabetes mellitus, anemia and deafness, due to mutations in *SLC19A2*, encoding a thiamine transporter protein [7]. Although long-term follow-up data are limited, supplementation with thiamine and insulin may improve the clinical symptoms.

CoQ_{10} defects: Ubiquinone (coenzyme Q_{10}, CoQ_{10}) deficiencies are heterogeneous groups of autosomal recessive conditions affecting both children and adults [8]. The increasing number of molecular defects in enzymes of the CoQ_{10} biosynthetic pathways (*PDSS1, PDSS2, COQ2, COQ4*, [9] *COQ6, COQ9, ADCK3*) underlies the importance of these conditions. Despite the identification of several primary CoQ_{10} deficiency genes, the number of reported patients is still low, and no true genotype-phenotype correlations are known, which makes the genetic diagnosis still difficult. In addition to primary CoQ_{10} deficiencies, where the mutation impairs a protein directly involved in CoQ_{10} biosynthesis, we can differentiate secondary deficiencies. CoQ_{10} supplementation may be beneficial in both primary and secondary deficiencies and therefore the early recognition of these diseases is of utmost importance [8].

Treatment of acute stroke in mitochondrial disease: Stroke-like episodes are hallmarks of several mitochondrial diseases, including MELAS. Based on the recommendations of the Mitochondrial Medicine Society (USA) [10], intravenous arginine hydrochloride can be considered in an acute setting of a stroke-like episode associated with the m.3243A>G mutation or in patients with stroke-like episodes in other genetic forms of mitochondrial disease. The efficiency of oral arginine supplementation to prevent strokes in MELAS syndrome needs further confirmation [10].

Pyruvate dehydrogenase complex (PDHc) deficiency: The pyruvate dehydrogenase complex (PDHc) is a mitochondrial matrix multi-enzyme complex that provides the link between glycolysis and the tricarboxylic acid (TCA) cycle presenting with neurological phenotypes and resulting in structural brain anomalies and epilepsy. Milder deficiency states present with variable manifestations that include cognitive delay, ataxia and seizures. The use of the ketogenic diet bypasses the metabolic block by providing a direct source of acetyl-CoA, leading to amelioration of some symptoms, such as seizures and lactic acidosis. However, a recent study in mitochondrial myopathy showed that the effect of a high-fat, low-carbohydrate diet (modified Atkins diet) may lead to selective damage of ragged red fibers and long-term improvement of muscle strength, suggesting activation of muscle regeneration [11]. Cofactor supplementations with thiamine, carnitine and lipoic acid have been also used in patients with PDHc deficiency, respectively, to control lactic acidosis and seizures with some success [10, 12].

Leber's hereditary optic neuropathy: A multicenter double-blind randomized placebo controlled trial (RHODOS) with Idebenone (900 mg/day) failed to show benefit for its primary endpoint; however, all of the secondary endpoints showed a positive trend toward visual improvement. Based on the positive results obtained in the follow-up analysis of the trial, in 2015, Idebenone received EMA approval for LHON; and can be prescribed in some countries. Also EPI-743 has been used with some initial benefit in a small, open labeled study [13].

Drug therapies beneficial in small studies: Multiple vitamins and cofactors are used in the treatment of mitochondrial disease, although these treatments are not standardized and wide variations exist. The most frequently used nutritional supplement is CoQ_{10} in its various forms (Ubiquinone, Ubiquinol, Idebenone); however, experimental evidence in randomized placebo controlled studies is missing. The Cochrane review of mitochondrial therapies found little evidence supporting the use of any vitamin or cofactor intervention [14]. In general CoQ_{10} can be offered for patients with mitochondrial disease, not only in primary CoQ_{10} biosynthesis deficiencies, and reduced CoQ_{10} (Ubiquinol) is the most bioavailable form. Alfa-lipoic acid, riboflavin, thiamine, folinic acid and biotin can be offered, especially if their deficiency has been shown as supporting therapy in periods of metabolic worsening. Carnitine should be used only if there is a documented deficiency and levels should be monitored during therapy.

Exercise: Exercise intolerance is a common feature of mitochondrial disease. However, mitochondrial patients usually benefit from exercise and should be encouraged to participate in tailored exercise programs [15]. Exercise-induced mitochondrial biogenesis is an appropriate target to improve function of patients with various types of mitochondrial disease. Endurance training improves peripheral muscle strength and results in increased mitochondrial content, maximal oxygen uptake and improved respiratory chain enzyme activities. Resistance exercise can increase muscle strength and recruitment of satellite cells in muscle fibers, which may have an effect on the heteroplasmy rate of mutations in the mtDNA.

Organ transplantation: Due to the strictly tissue-specific clinical presentation of some forms of mitochondrial disease, transplantation of the affected organs can be considered as a treatment of mitochondrial disease. Liver transplantation has been performed in patients with mutations in *DGUOK* and *POLG*. Although survival after liver transplantation in these patients is lower than survival in other indications, a significant proportion of patients benefits from liver transplantation with long-term survival and a stable neurological situation despite initial neurological abnormalities [16, 17]. Recently, liver transplantation seem to be beneficial in two very specific mitochondrial diseases, MNGIE and *ETHE1* mutations, where the common feature is ubiquitously (also in liver) expressed enzyme defect, and the clinical manifestations are caused by the accumulation of toxic metabolites [18, 19]. Allogenic bone marrow transplantation may also be considered in patients with MNGIE; however, severe complications and graft versus host reaction limits its applicability [20]. Heart transplantation has been beneficial in about 14 percent of patients with Barth syndrome (*TAZ* mutation) [21] and in a few patients with MELAS, and may be considered in severe, heart-specific manifestations of other mitochondrial disease.

Symptomatic therapies: Although very few causative therapies are available for mitochondrial disease, symptomatic treatments are often beneficial to ameliorate symptoms. In clinical practice, multiple supportive therapies have been postulated or used.

Neurological symptoms: Epilepsy is a common feature of mutations in the nuclear encoded polymerase gamma (*POLG*) and in mitochondrial tRNA genes (MELAS, MERRF). Interestingly, epilepsy seems less common in patients with isolated complex IV (cytochrome *c* oxidase, COX) deficiency (e.g., *SURF1* mutations). Early epileptic manifestations are frequent in patients with autosomal recessive *POLG* mutations. Introducing antiepileptic medication is of the utmost importance in these patients, because prolonged seizures may result in progressive neurological deterioration. Valproic acid treatment for epilepsy should be strictly contraindicated in POLG-related disease because of a profound liver toxicity. Some patients with autosomal dominant *POLG* mutations and rarely with mtDNA mutations may develop Parkinsonism, which is usually responsive on L-Dopa therapy. Ptosis is a very common symptom of different types of mitochondrial disease (Kearns-Sayre syndrome, CPEO). If severe, it should be operated on; otherwise it may result in visual impairment and headache.

Cardiac features: Cardiomyopathy is frequent and should be monitored and treated in mtDNA mutations (m.3243A>G, mt-tRNA^{Ile} mutations, m.8993 T>G/C) and in deficiencies of some nuclear genes (*MTO1, AARS2, SCO2, TAZ, SDHA*, etc.) that cause severe complex infantile or childhood-onset disease. The high risk of severe and fatal cardiac arrhythmias may necessitate the implantation of pacemakers or implantable defibrillators in Kearns-Sayre syndrome. In patients with severe, prominent cardiomyopathy, but less severe symptoms affecting the other organs, heart transplantation can be considered (see earlier in this chapter).

Liver features: Liver failure can be an isolated manifestation in mtDNA depletion due to mutations in the *DGUOK* gene. Liver transplantation may be considered if the neurological symptoms are not too severe (see earlier in this chapter). There is a severe, but reversible infantile liver failure due to *TRMU* mutations, which spontaneously recovers if the patients survive the first six months of age. Recognizing this condition and a very thorough monitoring of these patients is of the utmost importance and they should receive all life-sustaining measures, because they have a good prognosis with a complete or almost complete recovery.

Prevention of the transmission of mitochondrial disease: The identification of numerous novel nuclear disease genes in mitochondrial disease enables prenatal and preimplantation genetic diagnosis in several families. The prevention of the transmission of mtDNA mutations has developed rapidly, and prenatal and preimplantation genetic diagnosis have become an option for an increasing number of mtDNA mutations. New techniques that are based on in vitro fertilization (IVF), including pronuclear and metaphase II spindle transfer, have the potential to prevent the transmission of serious mtDNA diseases. In March 2015, mitochondrial donation (nuclear material from the biological parents and mitochondria from a donor oocyte) was approved by Parliament in the United Kingdom [22]. Based on better understanding of mtDNA replication in early-stage embryos, the regulatory authority in the United Kingdom gave the green light to "mitochondrial donation" techniques to reduce the risk of transmitting mitochondrial disorders [23] and it will be offered for patients to prevent the transmission of mtDNA mutations. The increasing number of options necessitates competent genetic counseling of patients with mitochondrial disease.

Therapies in the experimental phase: Gene replacement and RNA- or DNA-based therapies have to face more difficulties in mitochondrial disease, because of an additional barrier, the mitochondrial membrane. Numerous novel therapeutic targets have been proposed, including molecular manipulation of nuclear and mitochondrial DNA, shifting heteroplasmy, replacing dysfunctional protein, promoting neuroprotection and reducing maternal transmission of mutated mitochondrial DNA; of particular interest are stem cell therapies [1]. These approaches are in the preclinical experimental phase; however, clinical trials are approaching. AAV-mediated allotopic expression of the wild type human ND4 gene is currently being tested in LHON (m.11778A>G), administered by intraocular injection (GenSight). Clinical trials are planned in MNGIE with a liver-targeted AAV vector to replace TYMP in the liver, which would allow the elimination of toxic metabolites systemically [24]. Supplementation with nucleosides increases substrate availability and improves mtDNA copy numbers in a mouse model of *TK2* mutations [25]. It is a very promising treatment approach for patients with various forms of mtDNA depletion. Recently developed techniques in cell culture (TALEN, Crisp/Cas9, ZNF) aim to decrease mtDNA heteroplasmy. The rapid identification of potentially beneficial substrates based on either specific targets or high throughput screening is very promising, and currently several such substrates are being investigated for mitochondrial disease (see www.clinicaltrial.gov.com/) [26]. National and international patient cohorts and registries and improved outcome measures facilitate that more and better-designed clinical trials will be performed in larger numbers of patients with mitochondrial disease, enabling better evaluations of efficiency of these potential therapies [27].

Table 6.1 Summary of beneficial therapies that can be used in clinical practice

Riboflavin metabolism defects		
Clinical presentation	**Gene defect**	**Therapy**
Brown-Vialetto-Van Laere (BVVL) Fazio-Londe syndromes	SLC52A2, SLC52A3	High-dose oral riboflavin 10 mg/kg/day increased up to 50 mg/kg/day in children, 1,500 mg/day in adults
Multiple acyl-CoA dehydrogenase deficiency (MADD)	ETFDH (ETFA, ETFB)	Oral riboflavin 100–400 mg/day, (+CoQ$_{10}$) high glucose, diet, avoid fasting
Riboflavin responsive ACAD9-associated complex I deficiency	ACAD9	Oral riboflavin 100–400 mg/day
Other vitamin-responsive defects		
Biotin-responsive basal ganglia disease (BBGD)	SLC19A3	Oral biotin 1–2 mg/kg/day and thiamine 10–40 mg/kg/day
Thiamine-responsive megaloblastic anemia (TRMA)	SLC19A2	Oral thiamine 100–300 mg/day up to 10–40 mg/kg/day, insulin
Coenzyme Q$_{10}$ deficiencies		
Primary CoQ$_{10}$ deficiencies	PDSS1, PDSS2, COQ2, COQ4, COQ6, COQ9, ADCK3	High-dose oral CoQ$_{10}$ 10 mg/kg/day in children, up to 1,500 mg/day in adults
Secondary CoQ$_{10}$ deficiencies	APTX, ETFDH, ANO10	Oral CoQ$_{10}$ 100–500 mg/day
Mitochondrial disease	mutations in various mtDNA and nuclear genes	Oral CoQ$_{10}$ 100–300 mg/day
Treatment options in other mitochondrial disease		
Stroke-like episodes (MELAS)	m.3243A>G	i.v. arginine hydrochloride
Pyruvate dehydrogenase complex deficiency (PDHc)	PDHA, PDHB, DLAT, or other PDHc-related genes	Ketogenic diet, cofactor supplementation
Leber's hereditary optic neuropathy (LHON)	m.11778A>G, m.3460A>G, m.14484T>C	Oral Idebenone (900 mg/day) Oral EPI-743 (3 x 100–400 mg/day)
Exercise		
Any mitochondrial disease	Various mtDNA and nuclear genes	Endurance training, resistance training
Organ transplantation		
Mitochondrial liver failure	DGUOK, TRMU, POLG	Liver transplantation
Cardiomyopathy	TAZ, MELAS	Heart transplantation
Bone marrow	TYMP	Allogenic bone marrow transplantation
Blood cells	mtDNA deletion (Pearson) PUS1, YARS2 (MLASA)	Blood cell transfusion

References

1. Nightingale H, Pfeffer G, Bargiela D, et al. Emerging therapies for mitochondrial disorders. *Brain* 2016;139(Pt 6):1633–1648.

2. Foley AR, Menezes MP, Pandraud A, et al. Treatable childhood neuronopathy caused by mutations in riboflavin transporter RFVT2. *Brain* 2014;137 (Pt 1):44–56.

3. Horvath R. Update on clinical aspects and treatment of selected vitamin-responsive disorders II (riboflavin and CoQ 10). *J Inherit Metab Dis.* 2012;35(4):679–687.

4. Olsen RK, Koňaříková E, Giancaspero TA, et al. Riboflavin-responsive and non-responsive mutations in FAD synthase cause multiple Acyl-CoA Dehydrogenase and combined respiratory-chain deficiency. *Am J Hum Genet* 2016;98(6):1130–1145.

5. Schiff M, Haberberger B, Xia C, et al. Complex I assembly function and fatty acid oxidation enzyme activity of ACAD9 both contribute to disease severity in ACAD9 deficiency. *Hum Mol Genet* 2015;24(11):3238–3247.

6. Alfadhel M, Almuntashri M, Jadah RH, et al. Biotin-responsive basal ganglia disease should be renamed biotin-thiamine-responsive basal ganglia disease: A retrospective review of the clinical, radiological and molecular findings of 18 new cases. *Orphanet J Rare Dis.* 2013;8:83.

7. Bergmann AK, Sahai I, Falcone JF, et al. Thiamine-responsive megaloblastic anemia: Identification of novel compound heterozygotes and mutation update. *J Pediatr* 2009;155(6):888–892.e1.

8. Emmanuele V, López LC, Berardo A, et al. Heterogeneity of coenzyme Q_{10} deficiency: Patient study and literature review. *Arch Neurol* 2012;69(8):978–983.

9. Brea-Calvo G, Haack TB, Karall D, et al. COQ4 mutations cause a broad spectrum of mitochondrial disorders associated with CoQ_{10} deficiency. *Am J Hum Genet* 2015;96 (2):309–317.

10. Parikh S, Goldstein A, Koenig MK, et al. Diagnosis and management of mitochondrial disease: A consensus statement from the Mitochondrial Medicine Society. *Genet Med* 2015;17 (9):689–701.

11. Ahola S, Auranen M, Isohanni P, et al. Modified Atkins diet induces subacute selective ragged-red-fiber lysis in mitochondrial myopathy patients. *EMBO Mol Med* 2016;8(11):1234–1247.

12. Prasad C, Rupar T, Prasad AN. Pyruvate dehydrogenase deficiency and epilepsy. *Brain Dev* 2011;33(10):856–865.

13. Yu-Wai-Man P, Votruba M, Moore AT, Chinnery PF. Treatment strategies for inherited optic neuropathies: Past, present and future. *Eye (Lond)* 2014;28 (5):521–537.

14. Pfeffer G, Majamaa K, Turnbull DM, et al. Treatment for mitochondrial disorders. *Cochrane Database Syst Rev* 2012;4: CD004426.

15. Bates MG, Newman JH, Jakovljevic DG, et al. Defining cardiac adaptations and safety of endurance training in patients with m.3243A>G-related mitochondrial disease. *Int J Cardiol.* 2013;168 (4):3599–3608.

16. Grabhorn E, Tsiakas K, Herden U, et al. Long-term outcomes after liver transplantation for deoxyguanosine kinase deficiency: A single-center experience and a review of the literature. *Liver Transpl* 2014;20(4):464–472.

17. Hynynen J, Komulainen T, Tukiainen E, et al. Acute liver failure after valproate exposure in patients with POLG1 mutations and the prognosis after liver transplantation. *Liver Transpl* 2014;20(11):1402–1412.

18. De Giorgio R, Pironi L, Rinaldi R, et al. Liver transplantation for mitochondrial neurogastrointestinal encephalomyopathy. *Ann Neurol* 2016;80(3):448–455.

19. Dionisi-Vici C, Diodato D, Torre G, et al. Liver transplant in ethylmalonic encephalopathy: A new treatment for an otherwise fatal disease. *Brain* 2016;139 (Pt 4):1045–1051.

20. Halter J, Schüpbach WM, Casali C, et al. Allogeneic hematopoietic SCT as treatment option for patients with mitochondrial neurogastrointestinal encephalomyopathy (MNGIE): A consensus conference proposal for a standardized approach. *Bone Marrow Transplant* 2011;**46**(3):330–337.

21. Clarke SL, Bowron A, Gonzalez IL, et al. Barth syndrome. *Orphanet J Rare Dis* 2013;**8**:23.

22. Gorman GS, Grady JP, Ng Y, et al. Mitochondrial donation – how many women could benefit? *N Engl J Med* 2015;**372**(9):885–887.

23. Herbert M, Turnbull D. Mitochondrial donation – clearing the final regulatory hurdle in the United Kingdom. *N Engl J Med* 2017;**376**(2):171–173.

24. Torres-Torronteras J, Viscomi C, Cabrera-Pérez R, et al. Gene therapy using a liver-targeted AAV vector restores nucleoside and nucleotide homeostasis in a murine model of MNGIE. *Mol Ther* 2014;**22**(5):901–907.

25. Garone C, Garcia-Diaz B, Emmanuele V, et al. Deoxypyrimidine monophosphate bypass therapy for thymidine kinase 2 deficiency. *EMBO Mol Med* 2014;**6**(8):1016–1027.

26. Kanabus M, Heales SJ, Rahman S. Development of pharmacological strategies for mitochondrial disorders. *Br J Pharmacol* 2014;**171**(8):1798–1817.

27. Pfeffer G, Horvath R, Klopstock T, et al. New treatments for mitochondrial disease – no time to drop our standards. *Nat Rev Neurol* 2013;**9**(8):474–481.

Chapter 7

Neurology – the Central Nervous System

Michael J. Keogh, Hannah E. Steele and Patrick F. Chinnery

Introduction

Mitochondrial disorders present with clinical features affecting organs that are highly dependent upon adenosine triphosphate (ATP), and the central nervous system (CNS) is particularly vulnerable. Both localized dysfunction, occurring due to stroke-like episode, focal seizures or cerebellar ataxia, and global cerebral dysfunction, arising due to mitochondrial encephalopathy or generalized seizure disorder, are well described. The focal abnormalities can be acute and transient, or can evolve progressively over months and years – as seen in patients with basal ganglia disease causing dystonia. Likewise, the encephalopathy can be acute and fluctuate over time, possibly resolving completely, or may progress relentlessly over years or decades. Central neurological features are seen in both childhood and adult presentations, and often occur as part of a multisystem mitochondrial disorder with features affecting the peripheral nervous system or other organs and tissues. To make matters more complex, variable brain phenotypes can be seen in different members of the same family, particularly when there is an underlying pathogenic mitochondrial DNA (mtDNA) mutation with varied levels of heteroplasmy. Central neurological features often occur in conjunction with ocular or auditory involvement, which is considered in Chapters 9 and 10.

Clinical Presentation

The main central neurological features of mitochondrial disorders are shown in Table 7.1. These often form part of a discrete clinical syndrome, where additional features point to an underlying mitochondrial disorder. Alternatively, these features may occur in isolation, when the clinical presentation may resemble sporadic, degenerative or other neurogenetic disorders not primarily caused by mitochondrial dysfunction. Classical clinical syndromes are described later in this chapter and in Table 7.2, but some patients have only some of these features (so-called oligosymptomatic cases), and many present with an ill-defined clinical syndrome sharing few of these classical features.

Alpers syndrome: Alpers syndrome (also called the Alpers-Huttenlocher syndrome) typically presents in infancy or early childhood with intractable seizures following developmental delay, hypotonia and encephalomyopathy (see also Chapter 2). Multiple seizure types often coexist: focal seizures and myoclonus are frequent, with the progression of epilepsia partialis continua (EPC) to secondary generalized status epilepticus being characteristic. A key feature of Alpers syndrome is liver failure that, when the disorder presents in adult life, can be mild in the first instance. It is critically important to avoid sodium valproate as it can

Table 7.1 Central neurological features of mitochondrial disorders

Clinical feature	Investigations (results)	Treatment
Ataxia	MRI brain (cerebellar atrophy)	Symptomatic therapy
Dementia	MRI Brain (normal or atrophic) EEG (diffuse slowing) Psychometry (subcortical cognitive deficit)	Supportive therapy
Encephalopathy	MRI (nonspecific white matter or basal ganglia signal abnormalities) EEG (diffuse slowing)	Supportive therapy Avoidance of precipitating factors or medications
Migraine		Analgesics and/or tryptans; prophylactic therapy when appropriate
Movement disorder (Parkinsonism, dystonia or chorea)	Dopamine transporter SPECT (normal or reduced basal ganglia uptake)	Symptomatic
Seizures (generalized, focal and/or myoclonic)	EEG (epileptiform abnormality)	Anticonvulsants: as with other epilepsies, choice is largely dependent on whether seizures are focal or generalized. However, VPA is absolutely contraindicated in patients with *POLG*-related disorders)
Spasticity	CMCT (length-dependent slowing)	Symptomatic therapy
Stroke-like episodes	MRI (High-signal T2 abnormality not conforming to vascular territories, posterior-predominant)	L-arginine may be effective

CMCV = central motor conduction velocity, EEG = electroencephalogram, SPECT = single photon emission positron tomography, VPA = valproic acid

precipitate fatal liver failure in patients with Alpers syndrome due to autosomal recessive mutations in gene encoding the mtDNA polymerase γ, *POLG*.

Chronic progressive external ophthalmoplegia and Kearns-Sayre syndrome: Chronic progressive external ophthalmoplegia (CPEO) can be an isolated ocular myopathy, or form part of Kearns-Sayre syndrome (KSS). Kearns-Sayre syndrome typically presents with ophthalmoplegia and ptosis in teenage years or early adult life, and progresses to cause ataxia, encephalopathy, pigmentary retinopathy and cardiac conduction defects that may result in heart block (Figure 7.1). Sensorineural hearing loss, diabetes mellitus and myopathy are

Table 7.2 Classical mitochondrial clinical syndromes presenting with central neurological features in adults

Syndrome	Clinical symptoms/signs	Onset age	Genetics
Alpers-Huttenlocher syndrome, childhood myocerebral hepatopathy syndrome	Seizures, developmental delay, hypotonia, hepatic failure	Infancy/childhood/early adult life	Recessive mutations in *POLG*, or unknown for Alpers syndrome; mutations in *POLG*, *c10orf2*, *MPV17*, *DGUOK*.
Chronic progressive external ophthalmoplegia (CPEO)	Ptosis, ophthalmoparesis. Proximal myopathy often present. Various other clinical features variably present.	Any age of onset. Typically more severe phenotype with younger onset.	mtDNA single deletions mtDNA point mutations (including m.3243A>G, m.8344A>G) nDNA mutations (*POLG1*, *POLG2*, *SLC25A4*, *C10orf2*, *RRM2B*, *TK2*, *SPG7* and *OPA1*)
Kearns-Sayre syndrome (KSS)	CPEO, ptosis, pigmentary retinopathy, cardiac conduction abnormality, ataxia, CSF elevated protein, diabetes mellitus, sensorineural hearing loss, myopathy	<20 years	mtDNA single deletions
Leigh syndrome	Encephalopathy precipitated by illness, brainstem and cerebellar dysfunction, neuropathy, cardiomyopathy	Infancy. Occasionally in adult life	Recessive or X-linked mutations in nDNA-encoded respiratory chain components, and less commonly mtDNA point mutations (usually *MTATP6*)
Mitochondrial encephalopathy, lactic acidosis, stroke-like episodes (MELAS)	Stroke-like episodes with encephalopathy, migraine, seizures. Variable presence of myopathy, cardiomyopathy, deafness, endocrinopathy, ataxia. A minority of patients has CPEO.	Typically <40 years of age but childhood more common.	mtDNA point mutations (m.3243A>G in 80%, m.3256C>T, m.3271 T>C, m.4332 G>A, m.13513 G>A, m.13514A>G)

Table 7.2 (cont.)

Syndrome	Clinical symptoms/signs	Onset age	Genetics
Mitochondrial neurogastrointestinal encephalopathy (MNGIE)	CPEO, ptosis, GI dysmotility, proximal myopathy, axonal polyneuropathy, leukodystrophy.	Childhood to early adulthood	nDNA mutations in *TYMP*, MNGIE-like syndromes may occur due to nDNA gene mutations with CPEO
Myoclonus, epilepsy, and ragged red fibers (MERRF)	Stimulus sensitive myoclonus, generalized seizures, ataxia, cardiomyopathy. A minority of patients has CPEO.	Teenage or early adult life	mtDNA point mutations (m.8344A>G most common; m.8356 T>C, m.12147 G>A)
Neurogenic weakness with ataxia and retinitis pigmentosa (NARP)	Ataxia, pigmentary retinopathy, weakness	Childhood or early adult life	*MTATP6* mutation (usually at m.8993)
POLG-related Ataxia neuropathy syndromes (ANS): Including MIRAS, SCAE, SANDO, MEMSA	SANDO: CPEO, dysarthria, sensory neuropathy, cerebellar ataxia. Other ANS: Sensory axonal neuropathy with variable degrees of sensory and cerebellar ataxia. Epilepsy, dysarthria, or myopathy are present in some.	Teenage or adult life	nDNA mutations (*POLG*, *C10orf2*, *OPA1*)

MIRAS = mitochondrial recessive ataxia syndrome, SCAE = spinocerebellar ataxia with epilepsy, SANDO = sensory ataxia with dysphagia and ophthalmoplegia, MEMSA = myoclonic epilepsy myopathy sensory ataxia.

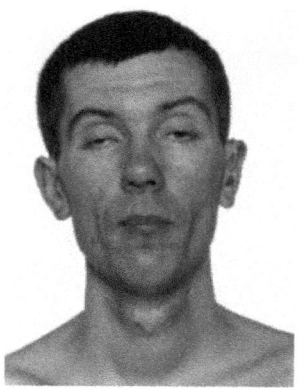 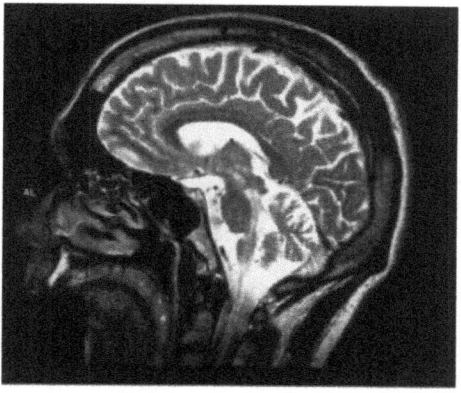

Figure 7.1 Clinical features and MR brain imaging in a patient with Kearns-Sayre syndrome caused by the 7kb deletion of mtDNA. Left – Ptosis and myopathic facies Right – Axial T2-weighted FSE MRI image showing generalized atrophy, and abnormal high signal in the midbrain and brainstem (adapted from *J Med Genet* 1999;36:425–426 & *Brain* 2000;123:82–92).

also common features. The usual cause is a single large-scale deletion in mtDNA and there is rarely a family history. The most important point to clinical management is regular surveillance for diabetes, and regular cardiac monitoring and treatment of heart block, which can cause unexplained sudden death.

Leigh syndrome: Although typically a childhood-onset disorder (see Chapter 2), the fluctuating brainstem dysfunction characteristic of Leigh syndrome with its associated neuropathy and extrapyramidal dysfunction can present in adulthood with an extrapyramidal disorder associated with cerebellar ataxia. Seizures occur in at least 40 percent of individuals, with more than half of those affected having a generalized seizure disorder characterized by myoclonic and/or absences. Focal seizures occur less frequently. The diagnosis is usually based on MR images showing typical basal ganglia lesions (Figure 7.2) and confirmed on biochemical or molecular genetic analysis.

Mitochondrial encephalomyopathy with lactic acidosis and stroke like episodes: Mitochondrial encephalomyopathy with lactic acidosis and stroke-like episodes (MELAS) can present at any age, but typically develops under the age of 40. Patients experience recurrent stroke-like episodes, often complicating migrainous headache. Seizures occur in around 60 percent of individuals with MELAS, which are characterized by prolonged focal seizures encompassing occipital lobe status, epilepsia partialis continuans (EPC) and other forms of non-convulsive status, as well as primary generalized seizures in a minority (Figure 7.3). Multisystem features include a cardiomyopathy, sensorineural deafness and diabetes. A minority of patients develop external ophthalmoplegia. In the early stages, the disorder can relapse and remit back to near normality, but with increasing frequency, the stroke-like episodes result in irreversible encephalopathy causing significant cognitive impairment. Anecdotally, the stroke-like episodes are more frequent in early life, and seem to "burn out" in middle age, resulting in a more slowly progressive neurodegenerative course. The m.3243A>G mitochondrial DNA mutation is the most common cause of MELAS, although several other point mutations and nuclear gene defects can cause

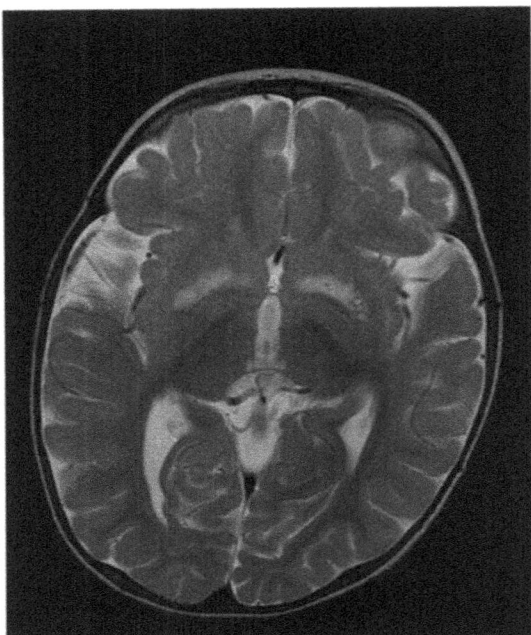

Figure 7.2 Bilateral, symmetric basal ganglia lesions on MR imaging in a patient with Leigh syndrome, in this case due to the m.9176 T>C mutation of mitochondrial DNA (adapted from *Pract Neurol.* 2015 Dec;15(6):424–435).

a similar clinical presentation. It is important to note that patients may have only one or two of the classical clinical features, and the full MELAS syndrome is actually uncommon in patients harboring the m.3243A>G mutation [1].

Mitochondrial neurogastrointestinal encephalomyopathy: Mitochondrial neurogastrointestinal encephalomyopathy (MNGIE) often presents in childhood or early adulthood with gastrointestinal features resulting in pseudo-obstruction or recurrent vomiting. Patients go on to develop ophthalmoplegia and ptosis, and a leukodystrophic encephalopathy (Figure 7.4). Peripheral neurological features include proximal myopathy and an axonal polyneuropathy. The disorder can be progressive, resulting in quadriplegia within years of onset. Patients have elevated thymidine levels detectable in plasma and an underlying recessive mutation in the gene-encoding thymidine phosphorylase (*TYMP*). Short-term benefits can be achieved with dialysis, but the most promising approach is bone marrow transplantation, which appears to arrest clinical progression [2]. A clinical trial is under way to evaluate this.

Myoclonic epilepsy with ragged red fibers: Myoclonic epilepsy with ragged red fibers (MERRF) is a progressive disorder developing in teenage or early adult life and is characterized by myoclonic epilepsy, generalized seizures and ataxia, in association with cognitive decline. Additional features include visual impairment due to an optic neuropathy, proximal myopathy and peripheral neuropathy. Cardiomyopathy is an important systemic feature. The most common cause is the m.8344A>G mtDNA mutation, although MERRF-like syndromes have been described in patients with other mtDNA and nuclear gene defects.

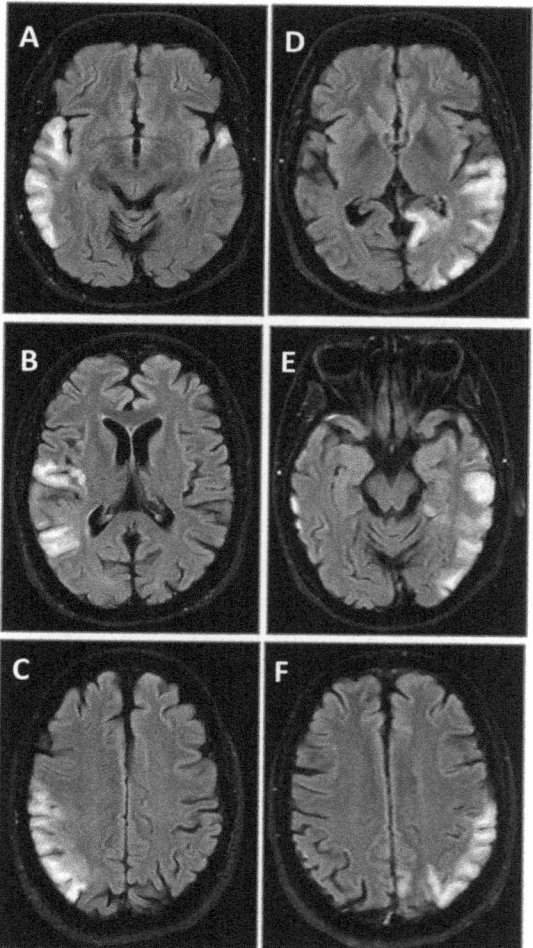

Figure 7.3 Serial MR imaging of two episodes in a patient with the m.3243A>G mutation. A–C, the first episode; D–F, the second. Note that the hyperintense changes in the temporal, parietal and occipital lobes do not conform to the vascular territories (adapted from *Pract Neurol.* 2015 Dec;15(6):424–435).

Neurogenic weakness with ataxia and retinitis pigmentosa: Neurogenic weakness with ataxia and retinitis pigmentosa (NARP) is a rare maternally inherited mitochondrial disorder caused by mutations in the *MTATP6* gene. The disorder typically presents in childhood or early adult life with ataxia and progressive visual failure due to the retinopathy, with associated proximal and distal weakness caused by myopathy and the peripheral neuropathy. The disorder is slowly progressive throughout adult life. Some family members develop Leigh syndrome and present at an earlier age (see Chapters 2 and 5) [3].

POLG-related ataxia and neuropathy syndromes: Autosomal recessive mutations in *POLG*, the gene encoding mtDNA polymerase γ, cause a group of overlapping clinical syndromes including mitochondrial recessive ataxic syndrome (MIRAS), spinocerebellar ataxia with epilepsy (SCAE), sensory ataxia with neuropathy, dysphagia and ophthalmoplegia (SANDO) and myoclonic epilepsy with myopathy and sensory ataxia (MEMSA). This

(a) (b)

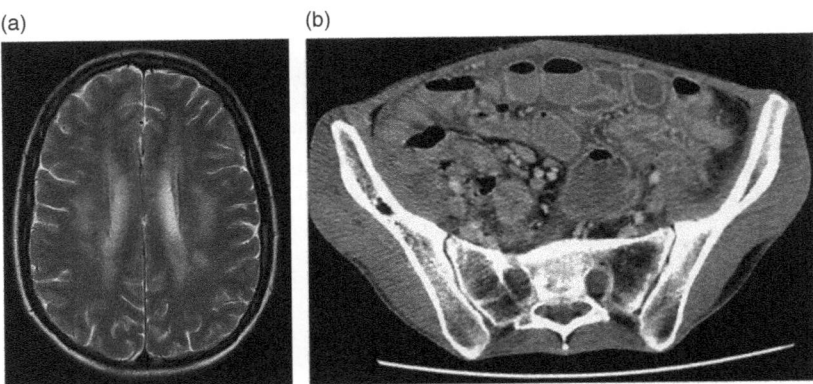

Figure 7.4 Mitochondrial neurogastrointestinal encephalomyopathy (MNGIE). The patient had an increase in the normally undetectable purines, thymidine and deoxyuridine, in both the plasma (13 and 17 μmol/l, respectively) and urine (0.2 and 0.3 mmol/l, respectively), very low white cell thymidine phosphorylase activity (2 nmol/h/mg protein; normal range: 336–1341) and compound heterozygous mutations in *TYMP* the TP gene (c.1167 T→C; c.1198_1203delGTGCTG). Left: Axial, T2-weighted MRI of the brain showing leukoencephalopathy. Right: intravenous contrast-enhanced CT of the abdomen and pelvis, revealing multiple loops of small bowel with fluid and gas distension but no transition point, suggestive of ileus/intestinal dysmotility (adapted from *Gut* 2011;60:805).

spectrum of disorders typically presents in teenage or early adult life, and ophthalmoplegia is associated with combined cerebellar and sensory ataxia [4]. The sensory ataxia component can be profoundly disabling and the disorder is relentlessly progressive. Recurrent stroke-like episodes resemble the MELAS syndrome (see earlier in this chapter), and epilepsy can be both a presenting feature (e.g., occipital lobe seizures) and an end-of-life event (e.g., intractable status epilepticus) [5]. As mentioned previously, sodium valproate should be avoided due to the risk of liver failure. Mutations in *POLG* cause secondary depletion and deletions of mtDNA, which lead to the clinical phenotype. Other so called mtDNA maintenance disorders can have similar presentations.

Clinical Examination

All patients with suspected mitochondrial disease should undergo a complete examination of the central nervous system, including an assessment of cognitive function. Cognitive defects can be mild, and often show a subcortical pattern, which can be picked up on neuropsychometric analysis. An assessment of consciousness may also be required in patients presenting acutely with encephalopathy or seizures. Careful observation is often helpful before commencing the formal examination, revealing myoclonic jerks or a movement disorder. Evaluating the gait is critically important, and is often the first clue to an underlying central neurological abnormality.

Cranial nerve examination can reveal visual field deficits due to retinopathy or optic neuropathy, which may be confirmed on fundoscopy (see Chapter 9). Visual field defects such as a centrocecal scotoma point to an optic neuropathy, whereas peripheral visual field defects could arise through stroke-like episodes seen in MELAS. Pupillary reflexes are often spared in mitochondrial optic neuropathies. Examining the eye movements can reveal an external ophthalmoplegia, a supranuclear eye movement disorder or nystagmus consistent with a cerebellar or brainstem lesion. Examination of the lower cranial nerves may reveal myopathic facies (Figure 7.1), sensorineural deafness or bulbar weakness, which is important to identify for clinical management.

Examination of the limbs may reveal generalized muscle wasting, either symmetric or asymmetric weakness, which can be proximal or distal depending on the underlying mechanism. There may be spasticity, although this is uncommon in mitochondrial disorders (notably present in patients with *SPG7* mutations) [6], but reflexes are often depressed due to a peripheral neuropathy or dorsal root ganglionopathy. Both midline and appendicular cerebellar syndromes are common.

Again it must be emphasized that patients may display all, some or none of these features, and that the clinical pattern will change over time and vary in different family members with the same genetic defect.

Clinical Investigations

Brain imaging: Brain imaging is mandatory in patients with clinical features suggestive of central neurological involvement [7]. Structural imaging can reveal focal or generalized atrophy, deep white matter lesions, focal basal ganglia abnormalities and stroke-like episodes. Gliosis may be apparent in advanced cases affecting the supratentorial, infratentorial regions and the spinal cord, which may be atrophic. The primary modality is magnetic resonance imaging, although CT can reveal basal ganglia calcification, which is usually bilateral. Functional MRI (fMRI) may reveal a reduction in N-acetyle aspartate associated with structural abnormalities, and proton spectroscopy may show an elevated lactate peak and/or an acidic pH which may be generalized or focal. Many of these features are nonspecific, although certain patterns are highly suggestive of a mitochondrial disorder, such as the characteristic neuroimaging described in patients with *POLG* disorders (Figure 7.5) and m.3243A>G with stroke-like episodes crossing vascular boundaries (Figure 7.3).

EEG Electroencephalograph: EEG should be considered in both the investigation of persons with suspected mitochondrial disorders and for individuals with known mitochondrial disease when presenting with stereotyped episodes, alterations in conscious level or behavioral change. The electroencephalograph (EEG) may show focal or generalized abnormalities consistent with underlying encephalopathy or seizure activity, and cortical myoclonus may be demonstrated in some patients. However, the most frequently identified abnormalities on inter-ictal EEG are nonspecific and include generalized slowing, multifocal discharges, focal discharges and generalized discharges. Although no single "diagnostic" EEG features exist, periodic lateralized epileptiform discharges (PLEDS) in children, a posterior predilection early in disease course or evidence of EEG progression should prompt consideration of a mitochondrial etiology.

Neuropsychometry: Formal neuropsychometric investigation often reveals subtle subcortical deficits in patients with a mitochondrial disorder. Dementia may develop in the late stages of the disease.

Central motor conduction velocities: central motor conduction velocities may be delayed in patients with mitochondrial disorders due to spinal cord lesions. The change is often length dependent [6].

It should be stressed that patients with central neurological involvement should also be investigated for peripheral neurological involvement, and have a formal ophthalmology

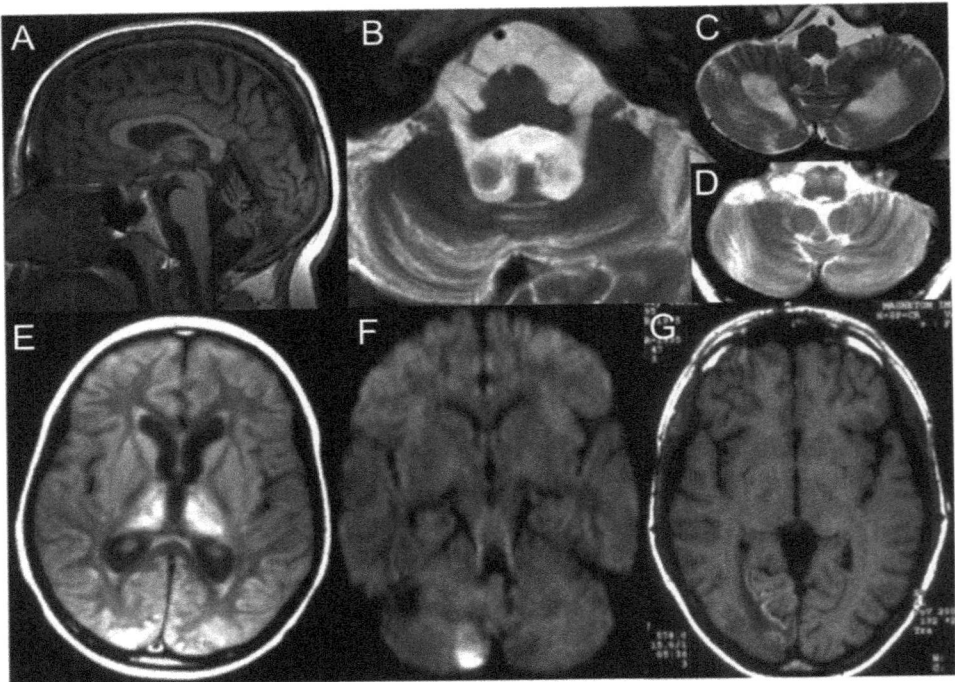

Figure 7.5 MRI findings in *POLG* encephalopathy
(A) Sagittal T1 image showing cerebellar atrophy (B) Axial T2 image showing dentate atrophy (arrow). (C) Axial T2 image showing cerebellar white matter hyperintensity. (D) Axial T2 image showing bilateral olivary lesions. The olives appear enlarged and hyperintense. (E) Axial T2 fluid-attenuated inversion recovery image showing bilateral thalamic and cortical occipital lesions. (F) Axial diffusion weighted image (b = 1000) showing acute stroke-like lesions in the right cerebellar cortex. (G) Axial T1 image showing linear, gyriform hyperintensity in the right medial occipital cortex (from *Brain* 2010;133:1428–1437).

assessment and audiogram. It is also important to remember extra neurological features including cardiomyopathy and diabetes.

Laboratory Investigations

Specific laboratory investigations have been considered earlier in this chapter, and are also shown in Table 7.2.

Clinical Management

The few notable exceptions (e.g., MNGIE, see Chapter 14) – there are no established disease-modifying therapies for central neurological features of mitochondrial disorders [8]. Treatment is largely supportive, using established approaches used in clinical neurology (see Chapter 6). These are discussed in Table 7.1. As mentioned earlier in this chapter, it is important to avoid sodium valproate in patients with suspected *POLG*-related disease or Alpers-Huttenlocher syndrome because of the risk of precipitating fatal liver failure. L-arginine has been proposed as a treatment for stroke-like episodes in m.3243A>G patients [9], but randomized trials have yet to be performed [8].

CASE REPORTS

Case Report 7.1

A 30-year-old woman presented to the Emergency Department with three unprovoked generalized tonic-clonic seizures within a 24-hour period. During her assessment, she had three further seizures that could not be terminated and required intravenous thiopentone and anesthesia. Brain imaging revealed bilateral occipital lesions. A detailed family history revealed that her mother died aged 54 from a cardiomyopathy, that her sister had bilateral sensorineural deafness and that another sister had attended the local diabetes clinic.

Self-Assessment Questions

- What is the likely diagnosis?
- What non-neurological investigations should be performed?
- What is the management plan?

Answers: The most likely molecular diagnosis is m.3243A>G. Although not noted at the time, the patient had short stature and had been complaining of hearing difficulties to her partner for about a year. The neuroimaging showed classical features of MELAS (Case Report 7.1 Figure), although the patient did not suffer from the full clinical syndrome. It transpired that the clinical problems affecting her other family members were also due to the same mitochondrial disorder, reflecting the varied oligo-symptomatic nature of the phenotype linked to differences in mtDNA heteroplasmy levels. The diagnosis was

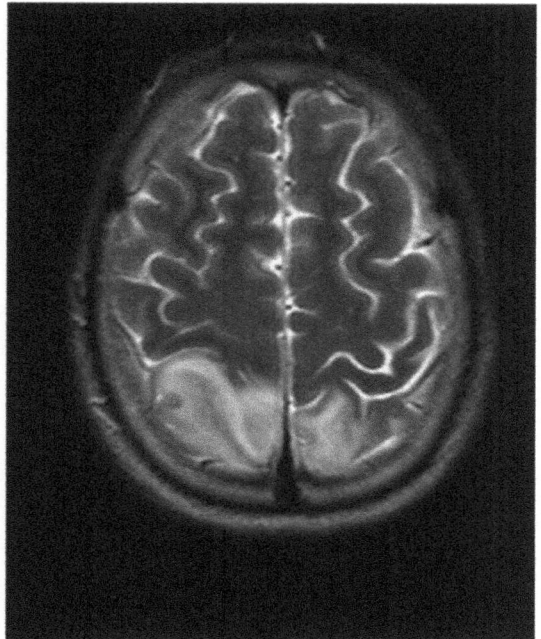

Case Report Figure 7.1

confirmed on muscle biopsy in this instance. The level of the mtDNA mutation is low in blood in adults, and can be undetectable using certain molecular techniques. It is therefore important to check the level of the mutation in urinary epithelial cells or in a muscle biopsy. Importantly, the muscle histochemistry may be entirely normal (although so-called strongly succinate dehydrogenase positive blood vessels have been described). Respiratory chain biochemistry can also be normal in patients with the m.3243A>G mutation, although defects of complex 1 and a combined defect involving a number of respiratory chain complexes are also described. It is important to check cardiac function with an ECG and an echocardiogram, and also to ensure the patient is not hyperglycemic on presentation.

Clinical management in this case was essentially supportive, the patient's seizures responded to therapy and she was discharged home. At a later stage, she required bilateral cochlear implantation for progressive sensorineural deafness. Over the subsequent decade, she developed significant progressive neurological disability linked to recurrent stroke-like episodes. Although she had no offspring, the m.3243A>G mutation is maternally inherited, and it is essential that she and other family members received appropriate genetic counseling.

Case Report 7.2

A 22-year-old woman, with no relevant family history of neurological disease, presented with jerking movements affecting the left hand. Initially thought to be functional, these persisted during her sleep and became disabling. The amplitude and severity varied over time, and there were periods when the focal jerking abated. After a normal brain CT and a normal EEG (between attacks) she was started on anticonvulsants, which improved the situation transiently. Unfortunately, the jerking movements recurred, and were intractable. She developed sensory ataxia over the subsequent three months and the seizures became secondary generalized. Although she was initially treated with lamotrigine and levetiracetam, escalating doses failed to control the jerking movements that persisted throughout the day and were associated with frequent generalized seizures. MR brain imaging was performed 18 months after her initial symptoms began.

Self-Assessment Questions

- What is the most likely diagnosis?
- What are the implications for her seizure management?
- What is the prognosis?

Answers: These are features consistent with mitochondrial recessive ataxia syndrome caused by autosomal recessive mutations in *POLG*. The patient was compound heterozygote with the two common mutations *A467 T* and *W748S*, which were confirmed to be in trans by testing the patient's parents who were heterozygote carriers for each allele. It is critically important to avoid sodium valproate use in patients with *POLG*-related disorders because of the risk of precipitating fulminant liver failure. The myoclonic jerks and epilepsia partialis continuans often respond well to levetiracetam in the short term, but eventually become intractable and progressive, leading to secondary generalized seizures, which can be aggressive and fatal.

Case Report 7.3

An 18-year-old man presented with an unsteady gait developing over six months. He also found it hard to hear his friends in public bars and struggled to read number plates in twilight. On examination he had a bilateral ptosis, which was not apparent on photographs taken during his early teenage years, and mild external ophthalmoplegia with no diplopia. Fundoscopy revealed specular pigmentary retinopathy. He had generally reduced muscle bulk, mild proximal muscle weakness and appendicular cerebellar ataxia. General medical examination revealed sinus bradycardia 38 beats per minute. A 12-lead ECG showed the following:

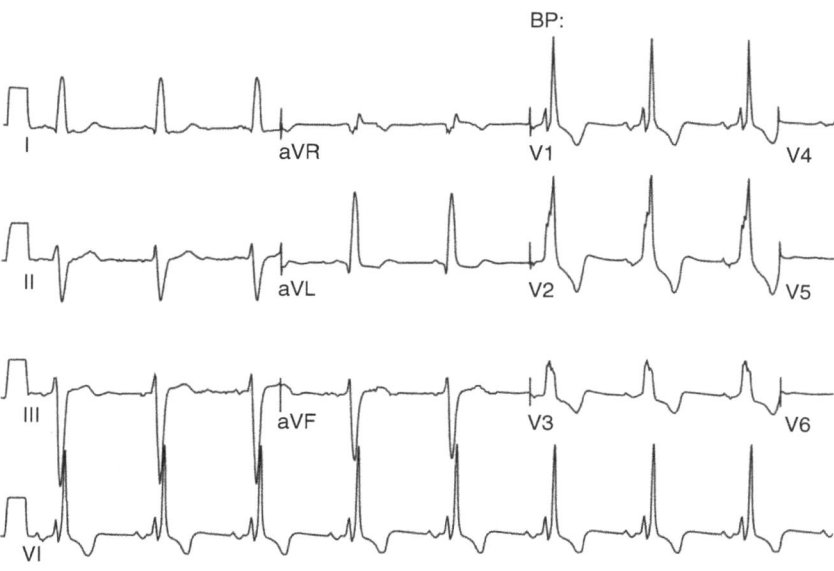

Self-Assessment Questions

- What is the likely clinical diagnosis?
- What investigations are required to confirm the diagnosis?
- What are the likely recurrence risks?

Answers: The most likely clinical diagnosis is Kearns-Sayre syndrome. This is most commonly caused by a single deletion of mitochondrial DNA that usually cannot be detected in blood. A muscle biopsy is, therefore, mandatory. This typically reveals a mosaic defect of cytochrome *c* oxidase and the presence of ragged red muscle fibers. Analysis of skeletal muscle mitochondrial DNA revealed a single 4.7 kilobase deletion of mitochondrial DNA in this patient at 72 percent heteroplasmy. The sinus bradycardia raised the possibility of heart block, which was confirmed on a 12-lead ECG in the clinic and led to acute cardiac pacing within 24 hours (bifasicular block in this case). Only on direct questioning did he admit to having had a number of fainting episodes in the previous months, most likely Stokes-Adams attacks due to episodes of extreme bradycardia. An oral glucose tolerance test revealed impaired glucose tolerance that was initially managed by dietary modification, but required insulin within three months. He went on to develop profound sensorineural hearing loss and dysphagia associated with profound disability related to his cerebellar ataxia.

Being caused by a single deletion of mtDNA, the clinical recurrence risks for Kearns-Sayre syndrome are very low. In women with single mtDNA deletions, the recurrence risks are approximately 1:24.

Case Report 7.4

A 60-year-old gentleman was referred by a neurologist to the genetic neurology clinic. He described progressive gait unsteadiness evolving over a decade. Prior to that, he had had a normal birth and development and was a keen sportsman. Work was becoming difficult in his role as a security officer because he found it increasingly challenging to walk around the site unaided. He had also noted minor clumsiness while using a computer keyboard. There was no family history.

On examination he had a broad-based ataxic gait. He had slurring dysarthria but normal cognitive function. Cranial nerve examination revealed normal visual fields and visual acuity but pale optic discs. He had restricted extraocular movements with jerky ocular pursuits, and gaze evoked nystagmus in the horizontal plane. The remaining cranial nerve examination was normal. Limb tone was increased in both legs. There was mild proximal muscle weakness in the lower limbs. Tendon reflexes were pathologically brisk throughout with spreading in the lower limbs and bilateral extensor plantar responses. There was mild limb dysmetria. There were no sensory signs and he was Romberg test negative.

Self-Assessment Questions

- What is the likely diagnosis?
- What will the muscle biopsy show?
- What is the clinical management?

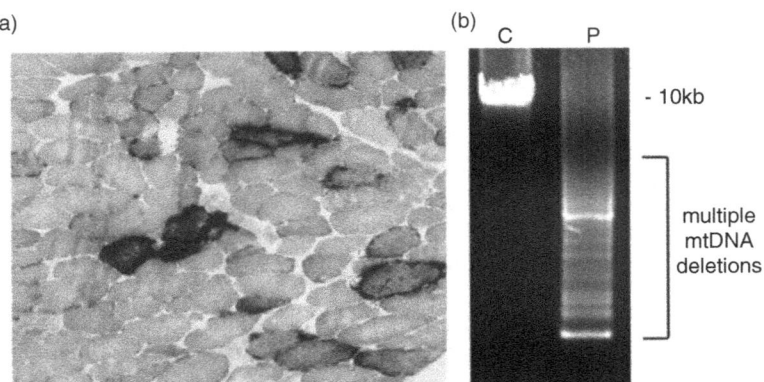

Case Report Figure 7.4 (A) Sequential COX/SDH histochemistry demonstrates a mosaic distribution of COX-deficient muscle fibers (blue) among fibers exhibiting normal COX activity (brown), with significant evidence of mitochondrial proliferation as shown by enhanced SDH reactivity around the subsarcolemmal region of the muscle fiber (ragged blue fibers). (B) Long-range PCR amplification of muscle DNA across the major arc shows significant evidence of multiple mtDNA deletions. C = Control; P = patient (adapted from *Brain* 2014 May;137(Pt. 5):1323–1336). (A black and white version of this figure will appear in some formats. For the color version, please refer to the plate section.)

Answers: The combination of a progressive ataxic syndrome with pyramidal signs in the absence of any family history suggests a diagnosis of autosomal recessive ataxia with spastic paraplegia. After excluding inflammatory causes on brain imaging, the most likely diagnosis is autosomal recessive hereditary spastic paraplegia caused by recessive mutations in *SPG7*. The lack of family history is typical in an outbred population and can be explained by high carrier frequency for specific recessive alleles.

The muscle biopsy often shows the presence of cytochrome *c* oxidase negative ragged red fibers at a higher frequency than expected due to normal aging (1–10 percent). Analysis of skeletal muscle mitochondrial DNA reveals the presence of multiple mtDNA deletions, which arise as a secondary phenomenon but may contribute to the phenotype when it includes ptosis and external ophthalmoplegia. Clinical management is largely supportive. Bladder dysfunction is common and urodynamic studies often reveal detrusor instability, which can be well managed with anticholinergic agents. The disorder is autosomal recessive, with corresponding recurrence risks.

References

1. Nesbitt V, Pitceathly RD, Turnbull DM, et al. The UK MRC Mitochondrial Disease Patient Cohort Study: Clinical phenotypes associated with the m.3243A>G mutation – implications for diagnosis and management. *J Neurol Neurosurg Psychiatry* 2013; **84**(8): 936–938.

2. Halter JP, Michael W, Schupbach M, et al. Allogeneic haematopoietic stem cell transplantation for mitochondrial neurogastrointestinal encephalomyopathy. *Brain* 2015.

3. Uziel G, Moroni I, Lamantea E, et al. Mitochondrial disease associated with the T8993 G mutation of the mitochondrial ATPase 6 gene: A clinical, biochemical, and molecular study in six families. *J Neurol Neurosurg Psychiatry* 1997; **63**(1): 16–22.

4. Horvath R, Hudson G, Ferrari G, et al. Phenotypic spectrum associated with mutations of the mitochondrial polymerase gamma gene. *Brain* 2006; 129(Pt. 7): 1674–1684.

5. Tzoulis C, Engelsen BA, Telstad W, et al. The spectrum of clinical disease caused by the A467 T and W748S POLG mutations: A study of 26 cases. *Brain* 2006.

6. Pfeffer G, Gorman GS, Griffin H, et al. Mutations in the SPG7 gene cause chronic progressive external ophthalmoplegia through disordered mitochondrial DNA maintenance. *Brain* 2014; **137**(Pt. 5): 1323–1336.

7. Bricout M, Grevent D, Lebre AS, et al. Brain imaging in mitochondrial respiratory chain deficiency: Combination of brain MRI features as a useful tool for genotype/phenotype correlations. *J Med Genet* 2014; **51**(7): 429–435.

8. Pfeffer G, Horvath R, Klopstock T, et al. New treatments for mitochondrial disease – no time to drop our standards. *Nature Reviews Neurology* 2013; **9**(8): 474–481.

9. Koga Y, Akita Y, Nishioka J, et al. L-arginine improves the symptoms of strokelike episodes in MELAS. *Neurology* 2005; **64**(4): 710–712.

Chapter

8 Neurology – the Peripheral Nervous System – Muscle and Nerve

Robert D. S. Pitceathly, Michael G. Hanna
and Mary M. Reilly

Introduction

Impaired mitochondrial function potentially disrupts a broad range of biological pathways and processes. However, despite the almost universal requirement of human cells for adenosine triphosphate (ATP), tissue-specific disease frequently occurs. When the molecular defect resides within the mitochondrial genome, the percentage of mutant versus wild-type mitochondrial DNA (mtDNA) molecules (so-called heteroplasmic mutant mtDNA load), tissue distribution (which can shift over time) and the precise timing of the mutational event (for example, during differentiation of the muscle precursor cells) may all contribute toward system-selective disease. The clinical heterogeneity associated with dysfunction of individual nuclear-encoded mitochondrial genes, and even identical mutations within those genes, is more difficult to explain. Epigenetic phenomena, genetic/environmental factors and the background mtDNA haplogroup have all been postulated as explanations for the phenotypic variability observed under such circumstances, but limited evidence exists for any of these hypotheses.

One factor that consistently contributes toward a given tissue's susceptibility to impaired oxidative phosphorylation (OXPHOS) is its metabolic demand. As such, the peripheral nervous system (peripheral nerve and skeletal muscle) is frequently implicated in both system-specific and multisystem disease, due to the exquisite sensitivity of neuronal and muscle tissue to reduced ATP levels. Muscle disease may manifest clinically as chronic progressive external ophthalmoplegia (CPEO) and proximal muscle weakness (which varies in severity) and with more nonspecific complaints, such as exercise intolerance, fatigue and myalgia. These symptoms can occur in isolation, as is often seen with single macrodeletions of mtDNA [1], or represent a single component of more complex multisystem disease, as is seen in Kearns-Sayre syndrome (KSS), myoclonic epilepsy with ragged red fibers (MERRF) and m.3243A>G-related mitochondrial disease. It should be emphasized that the latter does not necessarily conform to one of the well-recognized classical syndromic diagnoses, such as mitochondrial encephalomyopathy with lactic acidosis and stroke-like episodes (MELAS). Mitochondrial genetic testing should therefore be considered in those patients with coexisting muscle and central nervous system disease (including strokes, encephalopathy, migraine, seizures, myoclonus and ataxia) [2].

Although neuropathy as the sole presentation of mitochondrial disease is much less common than myopathy, a number of important exceptions exist. Such examples include mutations in *MFN2*, *GDAP1* and *MT-ATP6*, all of which encode mitochondrial proteins essential for normal peripheral nerve function and are discussed in further detail later in this chapter. A more frequent observation is the existence of neuropathy as part of a broader phenotypic spectrum, as seen with sensory ataxic neuropathy, dysarthria and ophthalmoparesis (SANDO) syndrome, which is usually caused by recessive *POLG* mutations.

This chapter focuses on the clinical presentation, examination findings, clinical/laboratory investigations and management of the peripheral nervous system sequelae that result from mitochondrial dysfunction. Two case studies that highlight important learning points are outlined.

Clinical Presentation

Adults with mitochondrial disease frequently present with neuromuscular symptoms [3] that include:

- CPEO, e.g., single mtDNA deletions, *MTTL1, MTTK, POLG, C10orf2, RRM2B*
- Myopathic proximal (common) muscle weakness, e.g., single mtDNA deletions
- Myopathic distal (rare) muscle weakness, e.g., *POLG*
- Neurogenic distal (common) muscle weakness, e.g., *MFN2, GDAP1, MTATP6/8*
- Neurogenic proximal (rare) muscle weakness, e.g., *TK2*
- Imbalance (sensory ataxia), e.g., *POLG, C10orf2, MTATP6*
- Dysarthria/dysphagia, e.g., single mtDNA deletions, *POLG, RRM2B*
- Exercise intolerance/fatigue/myalgia, e.g., *MTND2, MTND4, MTND5, MTCYB, MTCO2, MTTF, MTTY, MTTK, MTTE*
- Myoglobinuria/recurrent rhabdomyolysis, e.g., *MTCYB, MTCO1/2/3, MTTL1, DGUOK, HADHA/B*
- Periodic muscle weakness, e.g., *MTATP6/8*

Although many of these symptoms have a very heterogeneous genetic basis, a number of clinical clues help determine the underlying molecular defect. This is especially true for the multisystem diseases in which peripheral nerve and/or skeletal muscle pathology is just one component. For example, SANDO syndrome, most frequently due to recessive *POLG* mutations, causes a predominantly sensory neuropathy and, as its name suggests, a number of additional features, although dysarthria is not necessarily universal (and when absent the syndrome is termed sensory ataxic neuropathy and ophthalmoparesis [SANO] syndrome) [4]. Mitochondrial neurogastrointestinal encephalopathy (MNGIE) syndrome, a rare autosomal recessive disease associated with *TYMP* mutations, typically causes a prominent demyelinating peripheral neuropathy resembling chronic inflammatory demyelinating polyradiculoneuropathy (CIDP) or Charcot-Marie-Tooth (CMT) disease (pes cavus occurs in MNGIE). Furthermore, although numerous additional clinical sequelae accompany the peripheral nerve disease, including gastrointestinal dysmotility, cachexia, CPEO/ptosis and diffuse leukoencephalopathy, the neuropathy can be the initial manifestation [5].

Tissue-specific phenotypes are arguably more difficult to recognize as mitochondrial in origin. Primary neurogenic phenotypes include those caused by mutations in *MFN2* and *GDAP1*, nuclear-encoded mitochondrial genes necessary for effective fusion and fission of mitochondria (so-called mitochondrial dynamics). Autosomal dominant *MFN2* mutations are a common cause (20 percent) of axonal CMT (CMT2); they may be associated with upper motor neuron signs and optic atrophy, and frequently the neuropathy is more severe with an earlier onset than in other forms of CMT. *GDAP1* mutations usually cause axonal CMT (CMT2), although conduction velocities can be reduced into the CMT1 range, with the vast majority of cases inherited in an autosomal recessive pattern. The m.9185 T>C mutation in mitochondrially encoded ATP synthase 6 (*MTATP6*), an essential subunit of ATP synthase (complex V), can cause a pure motor/predominantly motor neuropathy (distal hereditary motor neuropathy/CMT2) [6]. Mutations in this gene can be easily overlooked if other family members are unaffected, and a matrilineal inheritance pattern is therefore not evident. This situation can arise if the mutation has arisen spontaneously in the affected individual or when the heteroplasmic mutant mtDNA level harbored by other relatives is below the threshold for disease to manifest. Additional complications related to this particular mutation may be present and should alert the clinician to the possibility of underlying mitochondrial dysfunction. These include learning difficulties, sensorineural hearing loss, cerebellar ataxia, retinitis pigmentosa, seizures, pyramidal tract signs and rapid clinical decompensation following intercurrent illness. Homoplasmic m.9185 T>C levels (i.e., 100 percent mutant mtDNA with undetectable levels of wild-type mtDNA) are associated with Leigh syndrome, a more severe phenotype of the central nervous system.

CPEO, the most frequent manifestation of myopathic mitochondrial disease, is usually caused by a single large-scale mtDNA deletion, but mtDNA point mutations and disruption to nuclear-encoded mitochondrial genes necessary for mtDNA replication and maintenance are also implicated. Although muscle disease is ubiquitous irrespective of the underlying genotype, the existence of an underlying peripheral polyneuropathy is highly predictive of a defect within the nuclear genome, in particular *POLG* [7]. Exercise intolerance, fatigue and myalgia are also common myopathic manifestations of mitochondrial disease and are associated with mutations within both mitochondrially encoded protein subunits and tRNA genes. As these symptoms can occur in isolation in patients with mitochondrial disease but are also extremely prevalent in the general population, confirmation of the underlying diagnosis frequently requires either a positive family history of maternally-inherited disease and/or a muscle biopsy.

A list of mitochondrial and nuclear-encoded genes that are often linked with peripheral nerve and/or skeletal muscle disease as either a major/presenting feature or as part of a broader multisystem disorder is outlined in Table 8.1.

Examination

A comprehensive neuromuscular assessment of any patient with suspected mitochondrial disease is essential and complements a detailed general neurological and systemic examination to ensure important non-neuromuscular features of mitochondrial dysfunction are not missed. The most frequent nerve- and muscle-related symptoms patients with mitochondrial disease describe are summarized earlier and should be elicited from the history. As with any neurological examination, the assessor should follow a structured procedure to avoid overlooking important clinical clues to the underlying diagnosis.

A myopathic gait indicates proximal pelvic girdle muscle weakness, while high steppage is generally caused by a neurogenic process affecting the muscles of the anterior compartments of the calves. Weakness of these muscle groups can also be demonstrated by asking the patient to rise from a seated position with arms crossed and by standing on heels.

Table 8.1 Mitochondrial- and nuclear-encoded genes associated with peripheral nerve and muscle disease as either a major/presenting complaint or as part of a multisystem disease

	Mitochondrial-encoded genes	Nuclear-encoded genes
A. Polyneuropathy as a major or presenting complaint	• Protein-encoding genes: *MTATP6, MTATP8*	*MFN2, GDAP1, DHTKD1, AIFM1, PDK3, C12orf65, POLG, C10orf2, TK2, TYMP, MPV17, SLC25A19, HADHA, HADHB, FRDA*
B. Multisystem disease associated with peripheral nerve involvement	• tRNA and rRNA genes: *MTTL1, MTTK, MTRNR1* • Protein-encoding genes: *MTND1, MTND4, MTND6, MTATP6, MTATP8* • Single large-scale deletions	*POLG, C10orf2, RRM2B, ANT1, TYMP, DGUOK, MPV17, OPA1, GFER, SUCLA2, SUCLG1, PDHA1, PDHC, SLC22A5, PDSS1, ADCK3, SURF1, COX10, ATP5E*
C. Myopathy as a major or presenting complaint	• tRNA and rRNA genes: *MTTF, MTTL1, MTTI, MTTQ, MTTM, MTTW, MTTA, MTTN, MTTC, MTTY, MTTS1, MTTD, MTTK, MTTR, MTTL2, MTTE, MTTT, MTTP, MTRNR2* • Protein-encoding genes: *MTND4, MTCYB, MTCO1, MTCO2, MTCO3, MTATP6, MTATP8* • Single large-scale deletions	*POLG, C10orf2, RRM2B, ANT1, TYMP, POLG2, TK2, DGUOK, ETFDH, MGME1, DNA2, SPG7, AFG3L2, HADHA, HADHB, ISCU, FDX1L*
D. Multisystem disease associated with skeletal muscle involvement	• tRNA and rRNA genes: *MTTF, MTTL1, MTTI, MTTQ, MTTW, MTTN, MTTC, MTTS1, MTTV, MTTD, MTTK, MTTG, MTTR, MTTH, MTTS2, MTTL2, MTTE, MTTT, MTTP, MTRNR2* • Protein-encoding genes: *MTND1, MTND4, MTND5, MTND6, MTCYB, MTCO1, MTCO2, MTCO3, MTATP6* • Single large-scale deletions	*POLG, C10orf2, RRM2B, TYMP, POLG2, TK2, DGUOK, MPV17, SUCLA2, SUCLG1, NDUFA11, NDUFAF1, COX6B1, COX10, SCO2, ATPAF2, PDSS2, COQ2, COQ4, TSFM, TUFM, NUBPL, YARS2, BOLA3, PDSS1, PUS1*

Romberg's test is positive when a large fiber sensory neuropathy or dorsal root gang-lionopathy exists.

Cranial nerve examination should record evidence of ptosis, ophthalmoparesis and facial/bulbar weakness, including assessment of neck flexors and extensors. Respiratory function must be recorded (and accompanies quantitative spirometry) by observing for abdominal paradox and cough/sniff strength documented.

Inspection of the trunk and limbs should account for any musculoskeletal deformities, such as kyphoscoliosis, scapula winging and pes cavus, in addition to the presence of ulceration, trophic skin changes, muscle wasting (including the distribution) and muscle fasciculations. An exaggerated lumbar lordosis implies weakness of axial muscles. Tone may be normal or reduced. Power usually conforms to the general principles that proximal muscle weakness is caused by an underlying myopathy, while distal muscle weakness is suggestive of peripheral nerve disease. However, there are reported cases of distal myopathic (*POLG*) and spinal muscular atrophy (SMA)-like (*TK2*) mitochondrial phenotypes, so these patterns should not be discounted as non-mitochondrial in origin. Reflex loss occurs when there is a ganglionopathy or polyneuropathy. Sensory examination may be abnormal. The pattern of sensory loss depends on the anatomical localization of the dysfunctional mitochondria within the peripheral nervous system. Such examples include a radiculopathy, ganglionopathy, plexopathy or polyneuropathy. The pattern of sensory loss to each modality (vibration, proprioception, pain and temperature perception) should be recorded.

Clinical Investigations

Neurophysiology should be considered as an extension of the clinical examination in any patient presenting with neuromuscular symptoms or signs. Nerve conduction studies (NCS) will confirm an underlying polyneuropathy and help distinguish demyelinating from axonal pathophysiological phenotypes. This information can be important in refining the most likely underlying mitochondrial subtype.

Needle electromyography (EMG) may demonstrate either neuropathic changes (spontaneous activity, such as fibrillations and positive sharp waves, indicative of acute denervation, and polyphasic, long-duration, high-amplitude, fast-firing motor units, suggestive of chronic denervation) or myopathic changes (spontaneous activity with short-duration, low-amplitude, rapidly recruiting motor units at low levels of force). However, unlike NCS, the EMG findings do not tend to contribute significantly toward prioritizing the most likely causative candidate gene.

Neuromuscular magnetic resonance imaging (MRI) is rapidly emerging as a powerful tool in delineating the pathological basis of acquired and inherited neuromuscular diseases. There is currently limited data correlating radiological appearances of muscles and nerves with the underlying genotype in mitochondrial disease. However, distal predominant fatty infiltration of upper and lower limb musculature with selective sparing of specific muscle groups has been reported in a patient with distal myopathy caused by a heterozygous *POLG* mutation [8]. Orbital MRI demonstrates extraocular muscle atrophy, characteristic signal changes and prolonged T2 in patients with CPEO associated with single mtDNA deletions [9].

Laboratory Investigations

Routine laboratory blood tests are often unhelpful during the diagnostic evaluation of patients with suspected mitochondrial disease. There may or may not be an elevated lactate and lactate/pyruvate ratio in serum; however, raised cerebrospinal fluid protein and lactate levels are a useful indicator of CNS disease. Creatine kinase (CK) is variably increased when myopathy is a prominent feature, but not usually above 1,000 IU/l. One exception is *TK2*-related mitochondrial disease in which CK may be elevated 5–10 times the upper limit of normal [10].

Skeletal muscle biopsy is often central to the diagnosis, particularly when genetic analysis has excluded the common mtDNA point mutations detectable in blood, such as m.3243A>G, m.8993C>T/G and m.8344A>G, and pathogenic variants in the most common nuclear-encoded mitochondrial candidate genes, for example, *POLG*. Muscle tissue should be examined histochemically for evidence of ragged red fibers and for succinate dehydrogenase/cytochrome *c* oxidase (COX, complex IV) deficiency. Ultrastructural examination can provide useful information concerning mitochondrial morphology. However, in our hands, it does not usually contribute significantly toward diagnosis when muscle histochemistry and respiratory chain enzyme analysis are available. Spectrophotometric assays of respiratory chain enzyme activities, particularly in children, may detect individual or combined enzymatic deficiencies that guide subsequent molecular genetic testing. Western blot analysis of one-dimensional blue-native polyacrylamide gels is available in some centers and enables analysis of the "native" protein structure of the respiratory chain complexes. This is especially useful when investigating genes encoding assembly factors that help construct the multimeric respiratory chain enzyme complexes. Finally, next-generation sequencing techniques are revolutionizing mtDNA (and nuclear DNA) analysis and are rapidly being introduced into diagnostic laboratories. However, extraction of mtDNA from muscle tissue is necessary in order to exclude the many pathogenic mtDNA point mutations, in addition to large-scale mtDNA deletions, which are undetectable in blood.

Nerve biopsy does not form part of the routine diagnostic work-up of patients with suspected mitochondrial disease and is seldom required in cases of presumed inherited neuropathy, unless there is a strong suspicion of an acquired, potentially treatable peripheral nerve disease. Usually the diagnosis of *MFN2-* and *GDAP1*-associated CMT is made either by Sanger sequencing of the relevant gene or as part of a multi-gene, next-generation sequencing neuropathy panel analysis.

Managing Complications

Management of mitochondrial diseases is primarily supportive and a number of options are available to help patients. Care is most effectively delivered in a multidisciplinary team setting that brings together and coordinates all the clinical specialists required. Currently no treatment is proven to slow or reverse its progression [11, 12].

Physiotherapy and tailored exercise help with exercise intolerance and fatigue in some patients. There is also evidence that suggests activating muscle satellite cells by specific exercise training regimens allows wild-type mtDNA to repopulate muscle tissue and effectively reduce mutant load below the biochemical threshold, which impairs OXPHOS [13]. Ptosis surgery is indicated in appropriate individuals with CPEO.

Patients with a significant neuropathy should receive advice regarding foot care and a podiatry review if at risk of skin ulceration. Referral to a physiotherapist and orthotist should be arranged if foot deformity and/or weakness is present. Conservative management includes stretching exercises, insoles or ankle foot orthoses. If these fail, surgical intervention aimed at improving pain and ambulation may be required.

Myalgia, muscle cramps and neuropathic pain can be treated with appropriate pharmacological agents. There is anecdotal evidence that co-enzyme Q_{10} (Ubiquinone) improves fatigue. However, randomized control trials have been unable to replicate this effect with statistical significance.

CASE REPORTS

Case Report 8.1

A 37-year-old woman of European descent born at term to non-consanguineous parents presented at 12 months with poor weight gain for which no cause was found. At 10 years, she had short stature and mild exercise intolerance. By 21 years, proximal muscle weakness and fatigue were prominent features and at 25 years, renal Fanconi syndrome and metabolic acidosis were detected. She subsequently developed premature ovarian failure. There was no family history of neuromuscular disease. Current exercise tolerance is 200 yards with a walking aid. Recent clinical examination showed short stature (height 4 ft. 10 in.) and a proximal myopathic gait, with upper and lower limb proximal muscle weakness (MRC grade 4/5) and absent tendon reflexes, but no other signs of neuropathy. There is an asymptomatic pigmentary maculopathy and mild right-sided sensorineural hearing loss. Plasma lactate was elevated (3.8 mmol/L) with a normal CSF lactate. Brain MRI was normal at 12 years. NCS revealed slowing of motor and sensory nerve conduction velocities. EMG demonstrated polyphasic units of brief duration in deltoid and iliopsoas.

Self-Assessment Questions

1. Consider the clinical phenotype described earlier in terms of tissue-specific versus multisystem disease, and whether there is a peripheral nerve involvement, myopathy or both. How would the peripheral nervous system involvement best be described in this case?
2. What investigations would be most appropriate at this point?
3. What would the most likely inheritance pattern be in this scenario?
4. What advice and management should be given to the patient?

Discussion

1) The patient has a multisystem disease with prominent muscle and, to a lesser extent, peripheral nerve involvement. Neurophysiology supports the clinical findings and is compatible with a (subclinical) demyelinating peripheral polyneuropathy and a proximal myopathy.
2) It would be appropriate to first exclude the common mtDNA point mutations detectable in blood (m.3243A>G, m.8993C>T/G and m.8344A>G). Pathogenic variants in *POLG* should also be considered given that premature ovarian failure is associated with mutations in this gene [14]. Genetic analysis for all of these variants was negative in this case. The next step would be to obtain muscle tissue for histochemical, biochemical and genetic analyses. Muscle biopsy revealed significantly decreased COX staining in the majority of fibers (Figure 8.1) with no ragged red fibers. Sural nerve biopsy was performed and showed moderate depletion of myelinated fibers with several large diameter axons with inappropriately thin myelin sheaths suggestive of demyelination (Figure 8.2, arrows). Spectrophotometric activity assay confirmed severe COX deficiency in skeletal muscle tissue, while respiratory chain complex I and complex II+III activities were within the reference range. Sanger sequencing of the entire mitochondrial genome was normal.

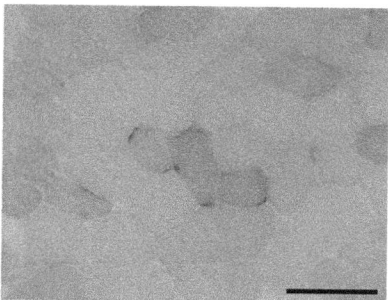

Figure 8.1 Muscle biopsy specimen shows decreased COX staining in the majority of fibers. Scale bars represent 100 μm. (A black and white version of this figure will appear in some formats. For the color version, please refer to the plate section.)

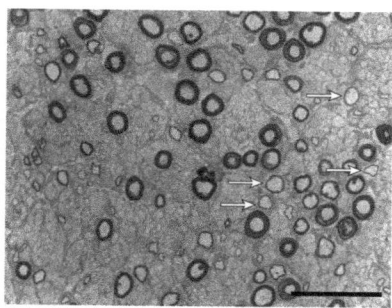

Figure 8.2 Sural nerve biopsy specimen. High-magnification view of one fascicle is shown (semithin section of resin stained with methylene blue-azure basic fuchsin). Multiple large-diameter axons with inappropriately thin myelin sheaths, suggesting demyelination, are shown by arrows. Scale bars represent 100 μm. (A black and white version of this figure will appear in some formats. For the color version, please refer to the plate section.)

3) The inheritance pattern is likely to be autosomal recessive given that both parents are unaffected (and assumed to be healthy carriers of the mutations). Furthermore, reduced enzymatic activity (in this case complex IV) is generally caused by homozygous or compound heterozygous "loss of function" mutations within the nuclear enzyme encoding genes. The latter is more likely in the case described as parents are non-consanguineous. Given the clinical phenotype (multisystem disease with prominent myopathic symptoms/signs and subclinical peripheral polyneuropathy) in the context of isolated COX deficiency, a complex IV polypeptide subunit or a complex IV assembly factor defect was considered. Compound heterozygous mutations within the nuclear-encoded complex IV assembly factor *COX10* were subsequently detected, thus confirming the genetic diagnosis in this case (see Table 8.1) [15].

4) Advice regarding foot care should be provided. As the neuropathy is currently subclinical, a podiatry and an orthotics review are not necessary at present. Physiotherapy and an exercise regimen might improve fatigue and exercise intolerance and co-enzyme Q_{10} could be administered.

Case Report 8.2

The index case (Figure 8.3 and Figure 8.4, III-8) presented with recurrent falls and foot drop aged six years following normal early development. He was subsequently diagnosed with mild learning difficulties and retinitis pigmentosa in adolescence. Examination aged 18 years showed distal muscle wasting of the legs, pes cavus and clawing of the toes (Figures 8.3A and 8.3B). Formal manual muscle strength testing (MRC graded) was normal

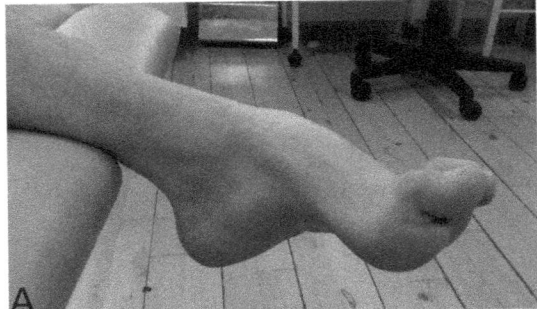

Figure 8.3 Photograph demonstrating distal lower limb muscle wasting, pes cavus and claw toes in index case (III-8, A and B).

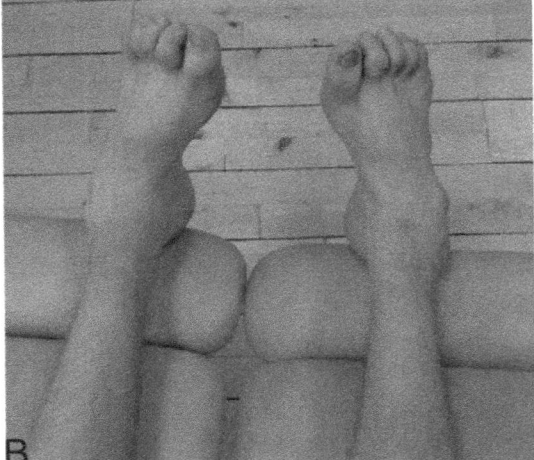

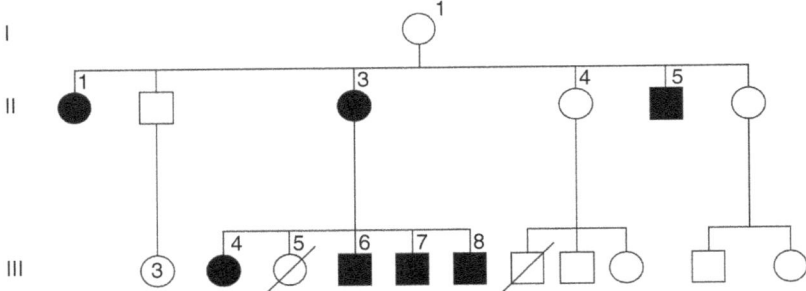

Figure 8.4 Pedigree. Filled symbols indicate individuals with axonal Charcot-Marie-Tooth type 2.

in the upper limbs with mild proximal lower limb weakness (hip flexion 4+/5 bilaterally) and moderate distal lower limb weakness (ankle dorsiflexion 3/5, plantarflexion 4+/5, inversion 5/5, and eversion 2/5 bilaterally). Ankle jerks were absent. Plantar responses were extensor. Sensory examination was initially entirely normal. However, when re-examined aged 21 years, vibration perception was reduced to both knees. Additional multisystem findings across the extended pedigree (Figure 8.4) included learning difficulties (II-3, II-5, III-4, III-5, III-6, III-7), sensorineural hearing loss (III-7) and cerebellar ataxia (II-3, III-4). CK, plasma and CSF lactate and brain MRI were normal. NCS aged 18 years were consistent with a pure motor axonal neuropathy.

Self-Assessment Questions

1. How would the peripheral nervous system involvement be best described in this case?
2. What inheritance pattern is the pedigree compatible with?
3. Which gene is most likely to present with this clinical phenotype?
4. What would the next appropriate investigative step be?
5. How should the patient be managed?

Discussion

1) The patient has a pure motor neuropathy, confirmed by NCS, as the predominant clinical feature, although there is a suggestion of multisystem involvement in the index case and the extended pedigree.

2) The inheritance pattern is consistent with maternal transmission, but would also be compatible with either autosomal dominant or X-linked inheritance. Although the initial clinical and neurophysiological phenotype is compatible with a distal hereditary motor neuropathy, the subsequent development of sensory signs is more in keeping with motor predominant axonal Charcot-Marie-Tooth disease type 2 (CMT2).

3) and 4) Given the associated multisystem involvement (especially retinitis pigmentosa and hearing loss) in the family and the potential matrilineal inheritance pattern, the underlying causative mutation most likely resides within *MTATP6* and the next appropriate investigative step would be to proceed directly to genetic analysis of this gene, which in this case confirmed the causative mutation m.9185 T>C [6]. Mutations within *MTATP6* are detectable in the blood of affected individuals. As such, muscle biopsy is seldom required for diagnostic purposes.

5) Vigilant foot care is still essential as sensory involvement may progress in later life. In view of the significant foot deformity and ankle weakness described, a physiotherapy and orthotics review would be appropriate. Surgical intervention to improve pain and ambulation might be necessary if conservative management fails.

References

1. Pitceathly RDS, Rahman S, Hanna MG. Single deletions in mitochondrial DNA – Molecular mechanisms and disease phenotypes in clinical practice. *Neuromuscul Disord* 2012 Jul;**22**(7):577–586.

2. Nesbitt V, Pitceathly RDS, Turnbull DM, et al. The UK MRC Mitochondrial Disease Patient Cohort Study: Clinical phenotypes associated with the m.3243A>G mutation – implications for diagnosis and management.

J Neurol Neurosurg Psychiatry. 2013 Aug;**84**(8):936–938.

3. Rahman S, Hanna MG. Diagnosis and therapy in neuromuscular disorders: Diagnosis and new treatments in mitochondrial diseases. *J Neurol Neurosurg Psychiatry* 2009 Sep;**80**(9):943–953.

4. Hanisch F, Kornhuber M, Alston CL, et al. SANDO syndrome in a cohort of 107 patients with CPEO and mitochondrial DNA deletions. *J Neurol Neurosurg Psychiatry* 2015 Jun;**86**(6)630–634.

5. Garone C, Tadesse S, Hirano M. Clinical and genetic spectrum of mitochondrial neurogastrointestinal encephalomyopathy. *Brain* 2011 Nov;**134**(Pt. 11):3326–3332.

6. Pitceathly RDS, Murphy SM, Cottenie E, et al. Genetic dysfunction of MT-ATP6 causes axonal Charcot-Marie-Tooth disease. *Neurology* 2012 Sep;**79**(11):1145–1154.

7. Horga A, Pitceathly RDS, Blake JC, et al. Peripheral neuropathy predicts nuclear gene defect in patients with mitochondrial ophthalmoplegia. *Brain* 2014 Dec;**137**(Pt. 12):3200–3212.

8. Pitceathly RDS, Tomlinson SE, Hargreaves I, et al. Distal myopathy with cachexia: An unrecognised phenotype caused by dominantly-inherited mitochondrial polymerase γ mutations. *J Neurol Neurosurg Psychiatry* 2013 Jan;**84**(1):107–110.

9. Pitceathly RDS, Morrow JM, Sinclair CDJ, Woodward C, Sweeney MG, Rahman S, et al. Extra-ocular muscle MRI in genetically-defined mitochondrial disease. *Eur Radiol* 2015 May 21. [Epub ahead of print]

10. Chanprasert S, Wong L-JC, Wang J, Scaglia F. TK2-related mitochondrial DNA depletion syndrome, myopathic form. In Pagon RA, Adam MP, Ardinger HH, Bird TD, Dolan CR, Fong C-T, et al., editors. GeneReviews(®) [Internet]. Seattle: University of Washington, Seattle; 1993–2015.

11. Pfeffer G, Majamaa K, Turnbull DM, et al. Treatment for mitochondrial disorders. *Cochrane Database Syst Rev Online* 2012 April;**4**:CD004426.

12. Pitceathly RDS, McFarland R. Mitochondrial myopathies in adults and children: Management and therapy development. *Curr Opin Neurol* 2014 Oct;**27**(5):576–582.

13. Clark KM, Bindoff LA, Lightowlers RN, et al. Reversal of a mitochondrial DNA defect in human skeletal muscle. *Nat Genet* 1997 Jul;**16**(3):222–224.

14. Duncan AJ, Knight JA, Costello H, et al. POLG mutations and age at menopause. *Hum Reprod Oxf Engl* 2012 Jul;**27**(7):2243–2244.

15. Pitceathly RDS, Taanman J-W, Rahman S, et al. COX10 mutations resulting in complex multisystem mitochondrial disease that remains stable into adulthood. *JAMA Neurol* 2013 Dec;**70**(12):1556–1561.

Ophthalmology

Patrick Yu-Wai-Man, Chiara La Morgia
and Valerio Carelli

Introduction

For most people, vision is the most precious of all the senses and on an evolutionary level, this importance is reflected by one third of the brain being devoted to handling visual information. The eye is a complex structure that allows the conversion of a light stimulus into an electrical signal that is then transmitted along the optic nerve and the retrochiasmal visual pathways toward the occipital cortex, where it is ultimately processed into conscious vision. As a paired organ, the extraocular muscles also play an important role in maintaining binocular single vision by aligning the eyes in different positions of gaze to achieve accurate foveation. Strikingly, more than half of all patients with confirmed mitochondrial disease develop sight-threatening complications that have a significant impact on quality of life and psychological well-being [1–2]. Ocular involvement can occur either in isolation or as part of a more complicated multisystem mitochondrial phenotype, further adding to the overall disease burden for patients and their carers. From a diagnostic point of view, the pattern of ocular tissues that are affected and the chronology involved can provide important clues, not only in raising the suspicion of an underlying mitochondrial cytopathy, but also by pointing toward the likely underlying molecular genetic defect. Despite major advances in our understanding of the disease mechanisms that precipitate mitochondrial eye diseases, patient counseling remains challenging due to the variable prognosis, which is frequently guarded, and the current lack of effective treatments. Nevertheless, the correct identification of an underlying ocular problem, if present, is an important element in the multidisciplinary management of patients with mitochondrial diseases as rehabilitative measures can have a positive impact on how they adapt and cope with their disabilities.

Clinical Presentations

Mitochondrial dysfunction in the eye has a particular predilection for retinal ganglion cells (RGCs), whose axons converge to form the optic nerve, and the extraocular muscles. Involvement of the optic nerve is a major cause of visual impairment and the two classical disease paradigms are Leber hereditary optic neuropathy (LHON) and autosomal dominant optic atrophy (DOA) [2, 3]. Both disorders result from the preferential loss of the small RGC axons within the papillomacular bundle that subserves central high-definition vision. LHON is characterized by subacute, bilateral, symmetrical visual loss predominantly in young adults, whereas DOA has a more insidious onset in early childhood. Significant recovery of visual function in LHON is rare and even if it does occur, it is incomplete, mostly limited to the appearance of areas of visual field fenestrations within the central scotoma. Although patients with DOA have a better visual prognosis compared with LHON, most of them are eventually registered as legally blind due to the progressive loss of RGCs and ongoing optic nerve neurodegeneration. The majority (50–70 percent) of patients with DOA harbor pathogenic mutations within the *OPA1* gene (3q28–q29), which encodes for an mitochondrial inner membrane protein (Table 9.1). LHON and DOA typically cause isolated optic neuropathies, but a subgroup of patients can develop a more severe syndromic "plus" form of the disease characterized by more widespread multisystemic involvement. These complications include a well-described multiple sclerosis-like illness in LHON (Harding disease) and varying combinations of chronic progressive external ophthalmoplegia(CPEO), sensorineural deafness, ataxia, myopathy and peripheral neuropathy in addition to optic atrophy in patients with DOA "plus" phenotypes [4, 5].

CPEO affects at least half of all patients with mitochondrial disease. Involvement of the extraocular muscles, which include the levator palpebrae superioris muscle, results in progressive ptosis and varying degrees of limitation of eye movements [6, 7]. The pattern of ophthalmoplegia is usually symmetrical, but about a third of all patients with CPEO will experience symptomatic diplopia that requires active management [8]. Although isolated cases can occur in the context of single large-scale mitochondrial DNA (mtDNA) deletions and milder nuclear genetic mutations, such as *PEO1*, CPEO is predominantly a multisystemic disease. Some frequently encountered clinical associations are still referred to by their eponymous appellations, although it has now been established that these are genetically heterogeneous. Classical CPEO syndromes include the Kearns-Sayre syndrome, which is a severe disease variant occurring before the age of 20 years old, and the syndrome of sensory ataxic neuropathy with dysarthria and ophthalmoparesis (SANDO), which is mostly seen in the context of mtDNA maintenance disorders secondary to *POLG* and *PEO1* mutations.

The retinal pigment epithelium (RPE) and photoreceptors within the outer retina can also be affected in a subgroup of patients with mitochondrial diseases, resulting in either a generalized pigmentary retinopathy or a more focal maculopathy. The pigmentary

Table 9.1 Nuclear mitochondrial disorders with prominent optic nerve involvement

Inheritance	Locus	*Gene*	OMIM	Phenotype
Dominant	1p36.2	*MFN2*	601152	Hereditary motor and sensory neuropathy type 6 (HMSN-6, CMT2A)
	3q28–q29	*OPA1*	165500	Isolated optic atrophy and syndromal dominant optic atrophy (DOA+)
	19q13.2–q13.3	*OPA3*	165300	Autosomal dominant optic atrophy and early-onset cataracts (ADOAC)
Recessive	5q.22.1	*SLC25A46*	616505	Optic atrophy and peripheral neuropathy
	9q13–q21.1	*FXN*	229300	Friedreich's ataxia (FRDA)
	11q14.1–q21	*TMEM126A*	612989	Optic atrophy ± auditory neuropathy
	16q24.3	*SPG7*	607259	Hereditary spastic paraplegia type 7 (HSP-7)
	19q13.2–q13.3	*OPA3*	258501	Type III 3-methylglutaconic aciduria (Costeff syndrome)

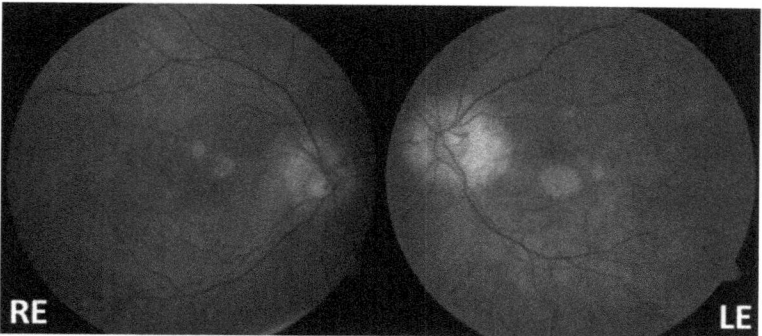

Figure 9.1 Pigmentary retinopathy in mitochondrial disease
There is generalized retinal pigment epithelial disturbance with areas of hypo- and hyper-pigmentation in this patient with Kearns-Sayre syndrome secondary to a single large-scale mtDNA deletion. A ring of peripapillary atrophy can also be observed, which is more prominent around the left optic disc. LE, left eye; RE, right eye. (A black and white version of this figure will appear in some formats. For the color version, please refer to the plate section.)

retinopathy may present as an isolated feature or more frequently, as part of a more extensive mitochondrial phenotype, for example, the Kearns-Sayre syndrome (Figure 9.1); the syndrome of neuropathy, ataxia and retinitis pigmentosa (NARP); or maternally inherited Leigh syndrome (MILS). Less frequently, some patients, especially in the pediatric group, can develop early-onset cataracts that may require surgical intervention depending on the impact of the lenticular opacities on visual function.

Clinical Examination

- The patient's visual acuity should be measured using a standard Snellen or LogMAR chart and a formal refraction is sometimes warranted to determine whether the subnormal vision is due to a refractive error, especially in children. Depending on the nature of the underlying ocular problem, the clinical examination can be focused to the relevant site of ocular pathology to determine the extent and severity of involvement, and to identify patterns that could point toward the underlying molecular diagnosis.
- The patient should be assessed for the presence of ptosis, which can sometimes be masked by compensatory frontalis overaction. In a nonspecialist setting, the severity of the ptosis can be roughly quantified as mild, moderate or severe, depending on the height of the lid margin in relation to the visual axis. For patients who might require surgical intervention, more detailed parameters need to be measured, in particular the margin reflex distance, levator function and frontalis power. The majority of patients with CPEO will demonstrate weakness of the orbicularis oculi muscle.
- The range of eye movements and the presence of any saccadic abnormalities need to be verified. If there are any limitations of the extraocular muscles, the degree of symmetry between the two eyes and the pattern of any strabismus, if present, should be documented. Patients with symptomatic diplopia should ideally be assessed by an experienced orthoptist to determine whether they would benefit from corrective prisms.
- The pupillary light reflexes and the presence of a relative afferent pupillary defect (RAPD) must be examined prior to instillation of any dilating drops. A careful neuro-ophthalmological examination will reveal additional features consistent with anterior

visual pathway involvement, namely, impaired colour vision and a dense central scotoma on confrontation field testing.

- A slit lamp examination is recommended to allow for a comprehensive review of the anterior and posterior segments of the eye. The presence of any diabetic retinopathy should be noted, although sight-threatening diseases are rare in patients with mitochondrial diabetes. Retinal pigmentary changes can be subtle and these can easily be missed if a dilated fundus examination of the macula and peripheral retina is not carried out.

Clinical Investigations

The onus is on the clinician to conduct a comprehensive set of investigations to delineate the nature of the underlying ocular problem in addition to seeking evidence of more disseminated systemic disease.

Ocular: A detailed slit lamp examination will provide important baseline findings that will help guide further investigations. The aim is to objectively assess various parameters, both structure and function, to give an indication of disease severity and to evaluate progression over time.

- Optical coherence tomography (OCT) is a noninvasive imaging modality that can provide high-resolution structural evaluation of the optic nerve and individual retinal layers, especially at the macula.
- A fluorescein angiogram can be useful to look for evidence of optic nerve leakage, macular oedema and vasculitis. In LHON, leakage is not observed at the optic nerve head, which is an important diagnostic clue in a patient presenting with an undifferentiated optic neuropathy.
- A baseline fundus photograph covering the area of pathology is recommended for future comparisons. Retinal autofluorescence can also provide useful information on the true extent of retinal pigment epithelial dysfunction and whether the central foveal region is involved.
- Formal assessment of visual fields with either static or kinetic perimetry should be carried out to determine the location, size and density of the patient's field defect. LHON is characterized by a dense central or centrocaecal scotoma, which can make accurate visual field perimetry challenging.
- Visual electrophysiology can be extremely helpful in determining whether visual loss is due to a primary RGC disease or outer retinal dysfunction and the evaluation should be done according to established ISCEV (International Society for Clinical Electrophysiology of Vision) standards. A standard protocol will include a full-field electroretinography (ERG), recordings of the P50 and N95 components of the pattern electroretinography (PERG) and measurements of the pattern and flash visual evoked potentials (VEPs). In primary optic neuropathies, VEPs show delayed latency and reduced amplitude of the cortical responses, and these responses can become undetectable in severe long-standing disease. The P50/N95 component of the PERG is a particularly sensitive measure of RGC damage, which can detect subtle abnormalities in the early stages of disease conversion in LHON [9].

Systemic: A broad list of differentials need to be kept in mind in the initial evaluation of a patient presenting with a rapid-onset unilateral or bilateral optic neuropathy, or a progressive bilateral ophthalmoplegia associated with variable degrees of ptosis. A detailed systemic evaluation should be conducted in conjunction with a neurologist and/or internist to look and more importantly, to exclude potentially treatable causes.

Neuroimaging: MRI of the brain and orbit with contrast is mandatory in all patients presenting with an as yet unexplained optic neuropathy. Atypical, enhancement of the optic nerve and chiasm has been reported in some patients with LHON [10]. In CPEO, the extraocular muscles are invariably globally atrophic and an orbital MRI series can be useful in the assessment of patients with ophthalmoplegia to exclude other more serious pathology [11].

Clinical Management

Some broad principles will be outlined that are relevant to most patients with mitochondrial eye diseases [1–3]. More specific management issues relevant to patients with LHON and CPEO will be discussed in greater detail in the two illustrative case reports provided later in this chapter. Patients with visual impairment should be reviewed, at least once, in a low visual aids clinic and the ophthalmologist coordinating their care should facilitate registration with their local social services. In some countries, eligible patients can be formally registered as legally blind or severely sight impaired, which provides them with additional benefits, including financial support. Children of different age groups will have particular needs and their care should be coordinated with the community paediatrician and the relevant social services to provide them with all the support that they need to remain in mainstream schooling if possible. Patients affected with mitochondrial eye diseases are often young adults who were previously fit and well, and the psychological impact cannot be underestimated. Occupational rehabilitation and close liaison with their employers can frequently allow these patients to retain their jobs, albeit with different roles to take into consideration their visual impairment. To prevent secondary complications from worsening the primary ocular pathology, patients should be screened for other systemic diseases, in particular diabetes mellitus, and if present, these should be actively managed with the help of an internist.

Common Conditions

LHON: Case Report 9.1 illustrates some of the key clinical features of LHON, namely the subacute progression of the optic neuropathy in this disorder, which is painless and invariably bilateral [1–3]. LHON is a disease of young adults with the peak age of onset being in the second and third decades of life. Three point mutations within the mitochondrial genome (m.3460 G>A, m.11778 G>A and m.14484 T>C) account for about 90 percent of cases, with the m.11778 G>A mutation being by far the most prevalent (60–80 percent) cause of LHON worldwide. Two peculiarities of this mitochondrial optic neuropathy are the marked incomplete penetrance and the male bias for visual loss. Although there can be wide intra- and interfamilial variability, the lifetime risk of visual loss is about 50 percent for male carriers and about 10 percent for female carriers, which reminds us that LHON should always be considered in all patients with an unexplained bilateral optic neuropathy, irrespective of sex. Access to genetic testing for the three primary LHON mtDNA mutations is now widely available and a confirmed molecular diagnosis can be rapidly established once LHON is suspected. The m.3460 G>A mutation is associated with the most severe biochemical phenotype and with the lowest likelihood of spontaneous visual recovery. In comparison, the m.14484 T>C mutation is the most benign and affected patients harbouring this mutation have the best visual prognosis, especially if disease conversion occurs before the age of 20 years old [12]. Unless there is a clear history of LHON in a family known to carry a disease causing mtDNA mutation, potentially more serious acquired optic neuropathies need to be considered and excluded.

The acute phase of LHON is dominated by a catastrophic wave of events that leads to rapid RGC death, axonal loss and optic atrophy. Two pathological observations with direct clinical correlates are worth emphasizing. First, there is swelling of RGC axons, which can be captured with OCT imaging as a thickening of the peripapillary retinal nerve fibre layer [13]. The latter proceeds in a specific chronological pattern that starts in the temporal and inferior quadrants, consistent with the preferential early involvement of the papillomacular bundle in LHON, followed subsequently by the superior quadrant and last by the nasal quadrant. Second, RGC loss is tightly related to axonal diameter and the axons that constitute the papillomacular bundle are particularly vulnerable due to their smaller size [14].

The loss of RGCs in LHON is due to mitochondrial dysfunction and the trigger is likely to be multifactorial with the underlying complex I respiratory chain defect impairing ATP production, and also increased chronic release of reactive oxygen species. Once a threshold is reached, which exceeds the cell's innate compensatory mechanisms, an irreversible cascade of events is initiated that precipitate apoptotic cell death. The incomplete penetrance observed in LHON reflects this precarious balancing act and the likelihood of a carrier losing vision is likely influenced by secondary genetic factors, both mitochondrial and nuclear, and environmental exposures. There is mounting evidence that smoking significantly increases the risk of disease conversion by further exacerbating the underlying mitochondrial bioenergetic deficit and unaffected carriers should, therefore, be strongly advised not to smoke or to phase out this habit [15].

Patient management: Genetic counseling for patients with LHON and their families is paramount to discuss the exclusive transmission of the pathogenic mtDNA LHON mutation along the maternal line; the risk of visual loss, which depends on the age and sex of the mutation carrier; and the need to avoid potential triggers of visual loss, in particular smoking and heavy drinking. Therapeutic options remain limited in LHON and management remains largely supportive with the provision of visual and occupational rehabilitation, and regular clinic follow-ups to document disease progression and any spontaneous visual recovery [17]. There is currently no prophylactic intervention that can prevent the onset of visual loss in at-risk LHON carriers or the involvement of the second unaffected eye after disease conversion. The only therapeutic option currently approved for affected patients with LHON is idebenone, which is a short-chain analogue of ubiquinone that has also been reported to improve mitochondrial ATP synthesis in addition to postulated antioxidant properties. Unlike coenzyme Q_{10}, idebenone can cross the blood–brain barrier, which probably accounts for its greater potency. Most published studies, including a multicenter double-blind randomized controlled trial, support a consistent visual benefit in a proportion of patients treated with idebenone, especially when started early during the acute phase of the disease process [18]. It must be stressed that idebenone will not completely reverse the significant damage already sustained to the optic nerve, but in those patients who respond to the drug, there is an increased rate and likelihood of visual recovery compared with the known natural history. Other therapeutic strategies under investigation for LHON include EPI-743, which is an α-tocotrienol-quinone derivative with antioxidant properties, and a gene therapy approach for the m.11778 G>A mtDNA mutation using the technique of allotopic gene expression and an intravitreal injection of a modified adeno-associated virus (AAV) vector to deliver the replacement *MTND4* subunit gene [17]. These interventions are not available outside the scope of clinical trials and it must be stressed that clear proof of efficacy in LHON has not yet been demonstrated.

Differential Diagnosis of LHON

Optic neuritis: Optic neuritis is the first disease manifestation in about 25 percent of patients who are eventually diagnosed with multiple sclerosis (MS). The peak age of onset is between 20 and 40 years old, and patients typically present with unilateral central loss of vision that worsens over two weeks before reaching the nadir. In the majority of cases (90 percent), there is associated ocular discomfort that is exacerbated by eye movements. There is an RAPD with marked dyschromatopsia and variable visual field defects. Demyelinating optic neuritis is usually retrobulbar, resulting in a normal looking optic disc, but in about a third of cases, there is optic disc swelling. Visual recovery starts within four to six weeks of disease onset and the prognosis is excellent with about 90 percent of patients recovering visual acuities of 6/12 or better. In most cases, an MRI scan performed in the acute phase will show optic nerve enhancement, but the main utility is for diagnostic and prognostic purposes as the presence of white matter lesions is a strong predictor of future progression to clinically definite MS.

Atypical optic neuritis: A number of red flags point toward a non-demyelinative etiology, including (i) bilateral simultaneous or rapidly sequential optic nerve involvement that is associated with poor visual recovery; (ii) marked optic disc swelling accompanied with retinal hemorrhages and/or exudates; and (iii) systemic features indicative of an underlying autoimmune tendency, such as lupus and Wegener granulomatosis. Neuromyelitis optica (NMO), in particular, should be considered in the differential diagnosis and aquaporin-4 (AQP4) and myelin oligodendrocyte glycoprotein (MOG) antibody testing should be requested in addition to a spinal cord MRI to look for longitudinally extensive transverse myelitis lesions. Infectious causes, such as syphilis, should also be ruled out as the visual loss is potentially reversible if appropriate treatment is initiated early.

Toxic-nutritional optic neuropathy: Patients may present at any age and the pattern of visual loss is usually bilateral, slowly progressive and painless. Colour vision is affected early and disproportionately to the observed visual acuity, and the visual fields typically show a caecocentral scotoma as in LHON and DOA. A careful history will reveal possible risk factors such as alcohol and tobacco abuse, a poor diet, chronic malabsorption states and exposure to drugs with well-reported mitochondrial toxic effects such as ethambutol and linezolid.

Infiltrative and compressive optic neuropathies: In a patient presenting with a rapidly evolving bilateral optic neuropathy, lymphoma and metastases need to be considered in the differential diagnosis. Compared with LHON, visual loss in DOA has a more indolent course, and it is imperative in this situation to exclude extrinsic compression of the optic nerve and/or chiasm. Two frequently missed diagnoses are optic nerve gliomas in children and meningiomas in middle-aged adults. An MRI scan of the brain and orbit with contrast and appropriate fat-suppression imaging is mandatory to avoid missing more subtle lesions involving the anterior visual pathways.

Case 2: CPEO: The extraocular muscles are involved in more than half of all patients with confirmed mitochondrial disease, either in isolation or as part of a more generalized systemic involvement. Ptosis is also the major feature for which patients seek medical attention in the first place due to the embarrassment felt in social circumstances. Interestingly, the severity of the extraocular muscle involvement in CPEO frequently overshadows the degree of skeletal muscle weakness, implying a differential susceptibility to the underlying mitochondrial dysfunction. Various hypotheses have been put forward to explain this tissue-specific vulnerability and some clues have emerged recently. Extraocular muscles are embryologically and genetically distinct from skeletal muscle with different anatomical and physiological properties [19]. These intrinsic differences are thought to predispose extraocular muscle fibres to accumulate secondary mtDNA abnormalities, such as multiple mtDNA deletions, at a much faster rate compared with skeletal muscle fibres [20]. Extraocular muscle fibres also seem to exhibit a lower threshold before an oxidative phosphorylative defect becomes apparent. In combination, these two factors could conceivably predispose extraocular muscle fibres to be more susceptible compared with skeletal muscle to an underlying genetic defect that impacts the normal biochemical functioning of the mitochondrial respiratory chain.

It has long been debated whether the limitation of eye movements in CPEO is due to a primary myopathic problem or whether there could be a superimposed supranuclear component. MRI studies have shown marked atrophy of the extraocular muscles in CPEO and formal eye movement recordings did not identify supranuclear dysfunction as a major contributory factor to ophthalmoplegia in the patients who were studied [11]. Pathological degeneration of the oculomotor nuclei within the brainstem supranuclear pathways could still play a role in some patients, but the current weight of evidence suggests that disease progression in CPEO is likely to derive mostly from the generalized atrophy of the extraocular muscles.

Age-related dehiscence of the levator aponeurosis, especially in current or previous contact lens wearers, is the most common underlying cause in individuals over the age of 50 years old. However, it is imperative for the clinician assessing the patient to have a mental checklist as a number of other oculomotility disorders can result in bilateral progressive ophthalmoplegia and ptosis, and differentiating these from CPEO is not always straightforward [6–8]. The three main diseases to consider are myasthenia gravis, oculopharyngeal muscular dystrophy and congenital cranial dysinnervation disorders.

Differential Diagnosis of CPEO

Myasthenia gravis: Myasthenia gravis has a predilection for the extraocular muscles and associated weakness of the orbicularis oculi muscle is a common finding, overlapping with some of the key features seen in CPEO. More than 50 percent of patients with myasthenia gravis will present with pure ocular features characterized by variable ptosis, fluctuating patterns of ophthalmoplegia and diplopia. Due to muscle fatigue, it is not uncommon for patients to report a diurnal pattern with their symptoms being worse in the evening. The rate of conversion of ocular myasthenia gravis into generalized systemic disease has been reported to be about 50 percent in the first two years after disease onset. Myasthenia gravis can mimic isolated cranial nerve palsies or present with more complex ocular motility patterns that evolve over time. Despite the multiple tests available in clinical practice, there is no reliable gold standard and making a definitive diagnosis can be challenging. Antiacetylcholinesterase antibodies are negative in 40–60 percent of cases. The ice pack test has gained increasingly popularity for assessing ptosis reversibility due to its simplicity, but it is not specific. Single-fibre EMG of the orbicularis oculi muscle has a relatively high diagnostic sensitivity and specificity, but it requires a skilled technician and it is only available in specialist centres. A positive response to a trial course of acetylcholinesterase inhibitor and/or steroids also strongly favors a diagnosis of myasthenia gravis. It is important to remember that the ophthalmoplegia in CPEO is generally bilateral, symmetrical and fixed, with none of the variable pattern seen with ocular myasthenia gravis.

Oculopharyngeal muscular dystrophy: The prevalence of oculopharyngeal muscular dystrophy (OPMD) has been estimated at 1 in 100,000, but the disease is more common in the French-Canadian population of Quebec and among Ashkenazi Jews. In the majority of cases, it is an autosomal dominant disorder caused by pathological GCG trinucleotide repeat expansions within the *PABPN1* gene, which encodes for the polyadenylate-binding nuclear protein 1. The autosomal recessive form of OPMD is relatively rare with only a few reported cases worldwide. Compared with CPEO, OPMD has an older age of onset from the fifth decade of life onward. Patients typically present with progressive ptosis, which is then followed by the development of dysphagia and proximal muscle weakness. Limitation of eye movements is observed in about 60 percent of patients, but complete ophthalmoplegia is rare and diplopia is uncommon. The clinical features of OPMD can develop insidiously and they can be variable. Compounded by the fact that it is a relatively rare disorder, diagnostic delays are frequent. A high index of clinical suspicion is therefore needed and all patients suspected of having CPEO should be specifically questioned about the presence of bulbar symptoms, in particular difficulties with swallowing and choking on solid foods.

Congenital cranial dysinnervation disorders: Congenital fibrosis of the extraocular muscles (CFEOM) is a genetically heterogeneous group of disorders characterized by nonprogressive ophthalmoplegia, which can be complicated with ptosis in some cases. Vertical gaze is usually more severely limited than horizontal gaze and patients frequently adopt an abnormal head position to compensate for the incomitant strabismus. So far, eight nuclear genes causing CFEOM phenotypes have been identified, and a new classification, under the label of congenital cranial dysinnervation disorders (CCDDs), has been proposed to highlight our improved understanding of the primary pathological process. Although congenital fibrosis of the extraocular muscles is the end-stage phenomenon that accounts for the observed ophthalmoplegia, all known CCDDs genes regulate critical pathways involved in oculomotor neurone development at the nuclear, brainstem or peripheral nerve levels, clearly implicating a primary neurogenic aetiology. The clinical presentation in early childhood and the nonprogressive nature of the ocular manifestations in CCDDs are usually sufficient pointers to allow a clear distinction to be made from classical CPEO.

CASE REPORTS

Case Report 9.1

A 16-year-old woman presented to her local ophthalmologist in October 2010 with rapid painless visual loss in her right eye. Her visual acuities were 6/12 in her right eye and 6/6 in her left eye. Colour vision was markedly reduced and static visual field perimetry showed a small, but dense central scotoma in the affected eye. The right optic disc was thought to be swollen and an urgent brain MRI scan with contrast was organized, which was reported as normal. When the patient was reviewed in February 2011, the vision in her right eye had deteriorated to 6/60 with an enlargement of the central scotoma. The patient was started on a 10-day course of treatment, initially with high-dose intravenous methylprednisolone and followed by a tapering course of oral prednisolone, but there was no improvement in visual function. Laboratory investigations, including syphilis, aquaporin-4 (AQP4) antibodies, folate and vitamin B12 levels, were normal. In March 2011, the patient was seen by another ophthalmologist and on closer questioning, a family history of blindness along the maternal line became evident. The patient's maternal grandmother was registered as legally blind after having suffered from rapid visual loss in both eyes at the age of 37 years old. There was some partial visual improvement one year after disease onset, but she remained severely visually impaired. A diagnosis of LHON was suspected and the patient was referred to a specialist neurogenetics clinic for further assessment.

Shortly after being referred, the patient started experiencing central visual blurring in her left eye and she was seen urgently. There was no RAPD and her visual acuities were counting fingers in her right eye and still 6/6 in her left eye (Figure 9.2, A and B). Optical coherence tomography revealed swelling in the peripapillary retinal nerve fibre layer, except

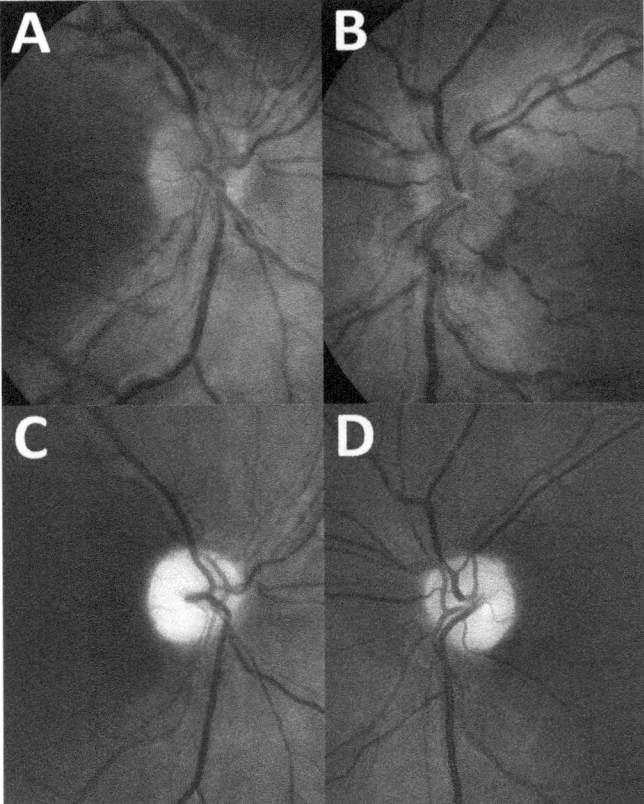

Figure 9.2 Optic disc appearance in the acute and chronic stages of LHON.
(A and B) In the acute stage, the patient's optic discs were hyperemic with peripapillary telangiectatic vessels and tortuosity of the central retinal vessels (March 2011). Mild temporal disc pallor was already evident in the right eye, and this was associated with retinal nerve fibre layer thinning within the papillomacular bundle on OCT. (C and D) The swelling of the retinal nerve fibre layer had resolved and both optic discs showed marked pallor of the neuroretinal rim (December 2013). (A black and white version of this figure will appear in some formats. For the color version, please refer to the plate section.)

in the temporal quadrant of the right eye, where diffuse thinning was noted. Molecular genetic investigation revealed that the patient was homoplasmic for the m.3460 G>A mtDNA mutation. The patient's vision in her left eye deteriorated rapidly over the next three months before stabilizing at counting fingers. At the last clinic visit in December 2013, her visual acuities were 6/60 in her right eye and 6/100 in her left eye (Figure 9.2, C and D).

Self-Assessment Questions

- What are the characteristic demographics of patients with LHON?
- What needs to be considered when taking a family history from someone with suspected LHON?
- What is frequently observed about the pupillary light reflex in LHON?
- Are the characteristic clinical signs always present in the acute stage?
- What additional ancillary tests may assist in the diagnosis?
- What are the diagnostic challenges when assessing a patient who is thought to be in chronic stage of LHON?

Answers

- LHON classically presents in young adult males with a peak age of onset in the second and third decades of life. However, this diagnosis should be considered in all cases of bilateral simultaneous or sequential optic neuropathy, irrespective of age and sex, in particular when visual function is severely affected with little or no recovery.
- There is frequently no clear history of other affected family members having experienced early-onset visual loss because of the reduced clinical penetrance masking the maternal inheritance pattern of LHON, or simply lack of extended family information.
- Relative preservation of the pupillary light reflex in the context of severe visual loss is a characteristic feature of LHON. This has been linked to the relative resistance of a special class of melanopsin retinal ganglion cells to neurodegeneration in mitochondrial optic neuropathies [16].
- In about 20 percent of cases of LHON, the fundus looks entirely normal in the acute stage and these patients, especially children, are frequently labeled as having functional visual loss. It should also be stressed that optic atrophy takes about six to eight weeks before it becomes clinically apparent.
- Useful ancillary tests in suspected cases of LHON are OCT, which can identify areas of segmental retinal nerve fibre layer swelling and thinning, and visual electrophysiology, which can provide objective evidence of RGC dysfunction [9, 13].
- If an individual is seen only in the chronic stage of LHON, it can be difficult to exclude other possible causes of optic atrophy, especially if there is no clear maternal family history. In these cases, MRI neuroimaging of the anterior visual pathways is mandatory while awaiting the results of the molecular genetic testing.

Diagnostic approach: A direct evaluation of the family history is crucial to establish whether there may be other affected family members, especially in more distant branches, and to clarify the possible mode of inheritance. A detailed neuro-ophthalmological examination and targeted ancillary investigations will exclude non-inherited causes (Sections 4 and 5). A lumbar puncture should be performed when clinically indicated to look for unmatched oligoclonal bands in suspected cases of demyelination or to exclude infectious and neoplastic causes. The appropriate neuroimaging modality should be requested and the films should ideally be reviewed with an experienced neuroradiologist. Mitochondrial genetic testing is the gold standard to confirm the presence of an underlying pathogenic mutation and to guide genetic counseling for the patient and the extended family. Most laboratories will screen for the three primary LHON mutations first (m.3460 G>A, m.11778 G>A and m.14484 T>C), which account for about 90 percent of all cases. If this is negative and there is a strong clinical suspicion, whole mitochondrial genome sequencing should be considered to identify rare mtDNA mutations known to cause LHON (Table 9.2).

Case Report 9.2

A 55-year-old man presented to his family physician complaining that his upper eyelids were drooping to the extent that he now had to lift them manually when watching television in the evening. The patient had type 2 diabetes mellitus and hypertension, both of which were well controlled with oral medication. He was a heavy smoker, but drank only socially.

Table 9.2 Mitochondrial DNA variants associated with LHON

	Mitochondrial Gene	Nucleotide Change
Common variants (~ 90%)	MTND1	m.3460 G>A*
	MTND4	m.11778 G>A*
	MTND6	m.14484 T>C*
Rare variants (~ 10%)	MTND1	m.3376 G>A, m.3635 G>A*, m.3697 G>A, m.3700 G>A*, m.3733 G>A*, m.4025C>T, m.4160 T>C, m.4171C>A*
	MTND2	m.4640C>A, m.5244 G>A
	MTND3	m.10237 T>C
	MTND4	m.11696 G>A, m.11253 T>C
	MTND4 L	m.10663 T>C*
	MTND5	m.12811 T>C, m.12848 C>T, m.13637A>G, m.13730 G>A
	MTND6	m.14325 T>C, m.14568 C>T, m.14459 G>A*, m.14729 G>A, m.14482 C>A*, m.14482 C>G*, m.14495A>G*, m.14498 C>T, m.14568 C>T*, m.14596A>T
	MTATP6	m.9101 T>C
	MTCO3	m.9804 G>A
	MTCYB	m.14831 G>A

* These mtDNA variants are definitely pathogenic. They have been identified in ≥ 2 independent LHON pedigrees and show segregation with affected disease status. The remaining putative LHON mutations have been found in singleton cases or in a single family, and additional evidence is required before pathogenicity can be irrefutably ascribed.

There was no history of ocular trauma and he had never worn contact lenses in the past. He denied any diplopia or dysphagia, and systemically, he felt well in himself. In the absence of any features suggestive of myasthenia gravis, his family physician referred the patient to his local eye unit for consideration of ptosis surgery. He was seen by a resident in the oculoplastic clinic who diagnosed involutional ptosis secondary to aponeurotic dehiscence of the levator palpebrae superioris. The patient was keen for surgical intervention and he was listed for a right anterior levator resection, to be followed by the left eye subsequently. On the day of the surgery, the attending surgeon noted reduced levator function bilaterally and limitation of eye movements in all cardinal positions of gaze (Figure 9.3). The surgical procedure was cancelled and the patient was referred to the neuro-ophthalmology clinic for an assessment of a possible myopathic ptosis.

On closer questioning, the patient had noticed drooping of his upper eyelids since his early 30s, but he had not been overly troubled by it until more recently. His parents and two older sisters did not have any eye problems, and he had two healthy children in their late 20s. Visual acuities were documented at 6/6 in both eyes. On examination, he had marked weakness of the orbicularis oculi muscle and levator function was reduced bilaterally (~ 7 millimeters). The patient had symmetrical limitation of eye movements and ocular saccades were slow, but there was no reported diplopia. There were no lens opacities and dilated fundus examination was normal. An MRI scan of the brain and orbit was organized. The extraocular muscles were globally atrophic and there was moderate cerebellar atrophy with some small vessel ischemic changes. The possibility of chronic progressive external ophthalmoplegia was considered in the differential diagnosis and a muscle biopsy was organized through the neurology service. Histochemical analysis showed 10–12 percent of cytochrome c oxidase (COX) negative muscle fibres and long-range PCR confirmed the presence of a single large-scale mtDNA deletion. The patient was still keen on surgical intervention and he underwent successful ptosis repair with a

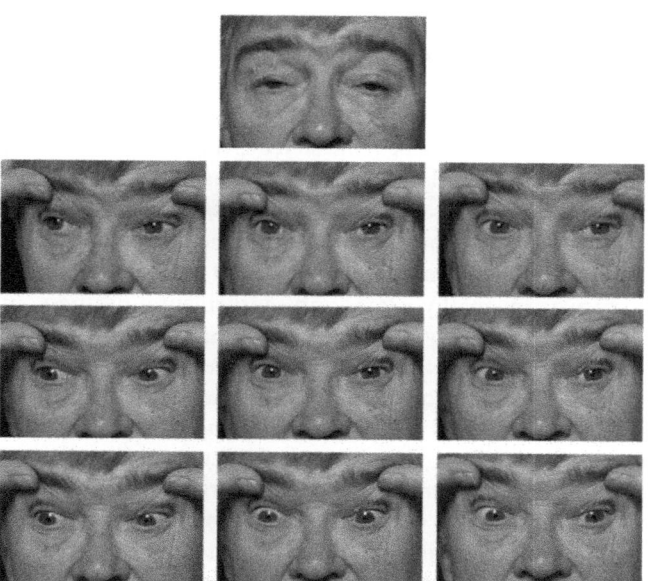

Figure 9.3 Chronic progressive external ophthamoplegia

The patient had significant bilateral ptosis and the upper eyelids were lifted to more clearly demonstrate the range of eye moments in the nine cardinal positions of gaze. There was symmetrical limitation of eye movements, which was worse on upgaze. (A black and white version of this figure will appear in some formats. For the color version, please refer to the plate section.)

more conservative anterior levator resection to minimize the risk of lagophthalmos and exposure keratopathy.

Self-Assessment Questions

- What clinical feature does patients with CPEO commonly exhibit?
- What pattern of eye movement abnormalities would be atypical in CPEO?
- What characteristic clinical feature should be sought when examining a patient with suspected CPEO?
- What saccadic eye movement abnormalities are usually observed in CPEO?
- What diagnosis should be considered if a patient with CPEO also has pigmentary retinopathy?
- What abnormalities are usually observed on MRI imaging of the orbits in patients with CPEO?

Answers

- Patients with CPEO almost always exhibit an exotropia (divergent strabismus).
- Unilateral or marked asymmetrical ophthalmoplegia are unusual in CPEO.
- The presence of orbicularis oculi weakness should be looked for specifically as it is a characteristic feature of CPEO.
- Even if formal eye movement recordings are not possible, the demonstration of slow ocular saccades on clinical examination is consistent with CPEO.
- Pigmentary retinopathy is an important hallmark of mitochondrial disease and it is part of the diagnostic triad for the Kearns-Sayre syndrome (Figure 9.1).
- Imaging of the orbit can be a useful adjunct in the diagnostic evaluation of patients with suspected CPEO as generalized atrophy of the extraocular muscles can be a striking morphological feature in this group of patients.

Diagnostic approach: Once a diagnosis of CPEO has been confirmed or is suspected, the patient should have a comprehensive ophthalmological and systemic workup. A dilated examination will reveal the presence of lens opacities, optic atrophy and/or retinal pigmentary changes. Other investigations such as visual electrophysiology, neuroimaging or tests to specifically exclude myasthenia gravis should be requested depending on the clinical scenario to rule out alternative pathologies. Ideally, the patient will have a full orthoptic assessment at baseline to document the dynamic range of eye movements, the presence of manifest or latent strabismus and the need for corrective prisms if there is symptomatic diplopia. Although CPEO can occur in isolation, more than 80 percent of all patients have other neurological deficits, in particular myopathy, ataxia and sensorineural deafness. In view of this high burden of associated morbidity, it is important for the patient to be reviewed by a neurologist and/or internist to allow for the prompt detection and management of related systemic complications. An electrocardiogram is also mandatory to identify possible cardiac conduction defects or evidence of ventricular hypertrophy.

A diagnostic muscle biopsy remains central to the evaluation of a patient with suspected CPEO to look for the characteristic pathological hallmarks of mitochondrial disease, namely a mosaic pattern of COX-negative muscle fibres and the presence of ragged red fibres. Long-range PCR on DNA extracted from the muscle biopsy specimen will determine whether the patient harbours a single large-scale mtDNA deletion, which remains the most common cause

Table 9.3 Nuclear genetic defects identified in patients with CPEO

	Autosomal dominant	Autosomal recessive
Multiple mtDNA deletions	POLG C10orf2 RRM2B OPA1 SPG7 ANT1	POLG POLG2 C10orf2 RRM2B SPG7 TK2 DGUOK MGME1
MtDNA depletion		POLG C10orf2 RRM2B TYMP

of CPEO. If multiple mtDNA deletions are detected instead, nuclear-encoded genes known to cause secondary mtDNA instability are then screened, starting with the more common ones such as POLG and PEO1 (Table 9.3). Targeted sequencing for specific pathogenic mtDNA point mutations, in particular the m.3243A>G mutation, should also be considered depending on the muscle biopsy findings and the patient's clinical features. The sequence in which these various genetic tests are performed will partly depend on the availability of a muscle biopsy service, ease of access to genetic testing and, importantly, cost considerations in some health care systems. Next-generation sequencing (NGS) technology, both exome and full genome, has led to a rapid acceleration in the identification of novel disease-causing genes in patients with clinically definite mitochondrial disease. The greater availability and affordability of NGS panels is likely to have a major impact on the diagnostic evaluation of patients with CPEO, allowing hopefully for a faster diagnosis, and with fewer genetically undetermined cases.

Diplopia

About a third of patients with CPEO will experience either intermittent or constant diplopia, which tends to be worse for near due to the eye's inability to converge properly. To determine the best management strategy, the angles of deviation should be measured at both near and distance fixation. The directions of gaze where diplopia are maximal should also be determined with a formal orthoptic assessment. In most cases, symptoms can be alleviated with the use of corrective prisms and these can be eventually be incorporated into the patient's glasses once a stable prism strength has been achieved. Surgery on the extraocular muscles in CPEO is rarely performed as it is a progressive disease and the strabismus invariably recurs. However, some patients are keen for surgical correction of a manifest strabismus for cosmetic reasons, more so when a large angle deviation is present. They should be appropriately counseled about the aims and limitations of strabismus surgery and if they do decide to go ahead, the two main options for improving ocular alignment are botulinum toxin injection and/or maximal strabismus surgery on the horizontal recti muscles.

Ptosis management: If severe, the ptosis in CPEO can lead to a visual field defect that impacts the patient's normal functioning. More importantly, patients can be particularly embarrassed about their facial appearance and the social impression that people who are unaware of their mitochondrial disease may have of them. Ptosis surgery can be highly effective in improving a patient's quality of life, but a conservative surgical approach is essential, and it should be carried out only by an experienced oculoplastic surgeon who can manage the possible complications effectively. Overcorrection of the ptosis in CPEO can lead to sight-threatening corneal exposure, which is magnified by the generalized ocular myopathy, including orbicularis oculi weakness, dry eye symptoms and a poor Bell's phenomenon. The surgical intervention needs to be carefully tailored by taking into consideration the residual function of the levator palpebrae superioris muscle, the power of the frontalis muscle and the patient's cosmetic preferences. The two most commonly performed ptosis procedures for CPEO are an anterior resection of the levator palpebrae superioris to maximize its upward muscle action, or if this is likely to be insufficient, a brow suspension can be performed that makes use of either a silicone sling or autologous fascia lata to transfer the mechanical strength afforded by the upward movement of the frontalis muscle onto the upper eyelid. Following successful surgery, the height of the upper eyelid can drop again due to the progressive nature of the disease and age-related changes. Revision surgery can be challenging and it is not always advisable if the risk-benefit ratio is not favourable, especially in terms of exposure keratopathy.

A confirmed molecular diagnosis of CPEO has obvious implications for other family members and appropriate genetic counseling should be provided, including specialist reproductive advice in some circumstances. The management of the more complex ocular symptoms in CPEO should ideally be overseen by an ophthalmologist with relevant knowledge and experience of mitochondrial disease. Patients with CPEO will frequently complain of dry eye symptoms due to a combination of lid margin disease (blepharitis) and insufficient wetting of the corneal surface by the tear film. Reassurance is key and patients or their carers should be encouraged to perform daily lid hygiene and when appropriate, use regular lubricating drops, which ideally should be preservative free to reduce the risk of corneal epithelial toxicity.

References

1. Fraser JA, Biousse V, Newman NJ. The neuro-ophthalmology of mitochondrial disease. *Surv Ophthalmol* 2010;**55**(4):299–334.

2. Yu-Wai-Man P, Griffiths PG, Chinnery PF. Mitochondrial optic neuropathies – disease mechanisms and therapeutic strategies. *Prog Retin Eye Res* 2011;**30**(2):81–114.

3. Yu-Wai-Man P, Chinnery PF. Leber hereditary optic neuropathy. In Pagon RA, Bird TC, Dolan CR, Stephens K, editors. Gene Reviews. 2013; available online at www.ncbi.nlm.nih.gov/books/NBK1174/ (Accessed December 8, 2015).

4. Pfeffer G, Burke A, Yu-Wai-Man P, Compston DA, Chinnery PF. Clinical features of MS associated with Leber hereditary optic neuropathy mtDNA mutations. *Neurology* 2013;**81**(24):2073–2081.

5. Yu-Wai-Man P, Griffiths PG, Gorman GS, et al. Multi-system neurological disease is common in patients with *OPA1* mutations. *Brain* 2010;**133**(3):771–786.

6. Bau V, Zierz S. Update on chronic progressive external ophthalmoplegia. *Strabismus* 2005;**13**(3):133–142.

7. Schoser BG, Pongratz D. Extraocular mitochondrial myopathies and their differential diagnoses. *Strabismus* 2006;**14**(2):107–113.

8. Richardson C, Smith T, Schaefer A, et al. Ocular motility findings in chronic progressive external ophthalmoplegia. *Eye* 2005;**19**(3):258–263.

9. Ziccardi L, Sadun F, De Negri AM, et al. Retinal function and neural conduction along the visual pathways in affected and unaffected carriers with Leber's hereditary optic neuropathy. *Invest Ophthalmol Vis Sci* 2013;**54**(10):6893–6901.

10. Phillips PH, Vaphiades M, Glasier CM, et al. Chiasmal enlargement and optic nerve enhancement on magnetic resonance imaging in Leber hereditary optic neuropathy. *Arch Ophthalmol* 2003;121(4):577–579.

11. Yu-Wai-Man C, Smith FE, Blamire A, et al. Extraocular muscle atrophy and central nervous system involvement in chronic progressive external ophthalmoplegia. *PLoS One.* 2013;8(9):e75048.

12. Barboni P, Savini G, Valentino ML, et al. Leber's hereditary optic neuropathy with childhood onset. *Invest Ophthalmol Vis Sci* 2006;47(12):5303–5309.

13. Barboni P, Savini G, Valentino ML, et al. Retinal nerve fiber layer evaluation by optical coherence tomography in Leber's hereditary optic neuropathy. *Ophthalmology* 2005;112(1):120–126.

14. Pan BX, Ross-Cisneros FN, Carelli V, et al. Mathematically modeling the involvement of axons in Leber's hereditary optic neuropathy. *Invest Ophthalmol Vis Sci* 2012;53(12):7608–7617.

15. Kirkman MA, Yu-Wai-Man P, Korsten A, et al. Gene-environment interactions in Leber hereditary optic neuropathy. *Brain* 2009;132(9):2317–2326.

16. La Morgia C, Ross-Cisneros FN, Sadun AA, et al. Melanopsin retinal ganglion cells are resistant to neurodegeneration in mitochondrial optic neuropathies. *Brain* 2010;133(8):2426–2438.

17. Yu-Wai-Man P, Votruba M, Moore AT, Chinnery PF. Treatment strategies for inherited optic neuropathies – past, present and future. *Eye* 2014;28(5):521–537.

18. Klopstock K, Yu-Wai-Man P, Dimitriadis K, et al. A randomized placebo-controlled trial of idebenone in Leber's hereditary optic neuropathy. *Brain* 2011;134 (9):2677–2686.

19. Yu Wai Man CY, Chinnery PF, Griffiths PG. Extraocular muscles have fundamentally distinct properties that make them selectively vulnerable to certain disorders. *Neuromuscul Disord* 2005;15(1):17–23.

20. Greaves LC, Yu-Wai-Man P, Blakely EL, et al. Mitochondrial DNA defects and selective extraocular muscle involvement in CPEO. *Invest Ophthalmol Vis Sci* 2010;51(7):3340–3346.

Chapter 10 Audiology

Peter Kullar

Introduction

Hearing depends on the mechanotransduction of sound pressure waves by the inner hair cells of the cochlea into neural signals relayed by auditory neurons to the auditory cortex. Dysfunction in any part of this process results in hearing impairment, a dynamic, multifactorial condition prevalent both in patients with mitochondrial disease and the general population. Epidemiological studies have indicated that approximately 1 in 500 at birth and more than 70 percent of those over the age of 70 are affected by some degree of hearing loss [1]. In those affected at birth, it is estimated that more than 50 percent have a genetic etiology with either an autosomal recessive, autosomal dominant, X-linked or mitochondrial mode of inheritance. Mitochondrial disorders contribute significantly to cases of genetic hearing loss and are estimated to be causative in approximately 5 percent of non-syndromic postlingual hearing loss as well as 1 percent of prelingual cases [2, 3]. Studies have also shown that mitochondrial mutations may also contribute to age-associated hearing loss [4].

Anecdotally, hearing loss is a well-recognized feature of mitochondrial disease, although ascertaining its prevalence in patients with confirmed mitochondrial disease is complicated by both clinical and genetic heterogeneity. Additionally, the true burden of disease may be further disguised by under-diagnosis, particularly in patients with complex neuromuscular phenotypes. In one study of 23 patients with mitochondrial disease, including 10 with the m.3243A>G mutation, 74 percent were found to have hearing loss [5]. Additional studies investigating patients with a range of mitochondrial diseases have further confirmed the high rate of hearing impairment in patients with mitochondrial disease [6, 7].

Mitochondrial dysfunction can be causative for hearing loss both in isolation (non-syndromic) and as a feature of systemic mitochondrial disease (syndromic). It has been shown that the cells of the auditory sensory axis, including the cochlea hair cells, stria vascularis (primarily responsible for maintaining the endocochlear potential) and auditory neurons, are highly metabolically active and therefore contain an abundance of mitochondria. A number of studies have implicated the cochlea as the primary origin of disease with a corresponding loss of both outer and inner hair cells [5, 8]. However, auditory neuropathy (i.e., disordered hearing with preserved cochlear function) has also been recognized as an important factor in a subset of mitochondrial diseases [9 10]. The sensitivity of the auditory pathway to mitochondrial dysfunction may result from inadequate mitochondrial oxidative phosphorylation in these metabolically active tissues. To date, the underlying molecular mechanisms, including those governing tissue specificity, pose a challenging and broadly unresolved question.

Clinical Presentation

The hearing loss of mitochondrial disease is sensorineural, symmetrical and primarily affects the higher frequencies, although progressive disease can lead to pan-frequency hearing loss (hearing loss at all tested frequencies). The incidence of conductive hearing loss, where sound conduction to the cochlea is reduced by disease of the middle or outer ears, for example in otitis media with effusion, has been shown to be comparable to the general population [11].

Hearing loss onset tends to occur in early life and to be gradually progressive, although sudden hearing loss has been described in stroke-like episodes in patients carrying the m.3243A>G mutation [12]. Predominantly, hearing loss occurs in later childhood or adulthood after the acquisition of speech. However, broad phenotypic variation means there often exists a range of hearing loss onset and severity even in family members carrying the same genetic variants. It has also been shown that hearing thresholds decline at a faster rate in m.3243A>G mutation carriers compared to the general population, meaning it is important that patients have regular audiology assessment [13].

Clinical Diagnosis

A mitochondrial cause for hearing loss may be suspected in cases that appear to show maternal transmission or where hearing loss is a known association of systemic mitochondrial disease. There are two main subgroups of patients presenting with potential mitochondrial associated deafness:

- Children with hearing loss of unknown origin (who may or may not have passed newborn hearing screening due to the variability in onset of hearing loss).
- Adults with progressive hearing loss with or without a known background of mitochondrial disease.

Although management of the hearing loss is the primary focus of clinicians, fully understanding the etiology is important for a number of reasons:

- Risk of recurrence in future offspring;
- Risk of progression (natural history);
- To exclude syndromic forms of deafness where associated symptoms may require further management.

In both children and adults, a detailed history should be undertaken to determine the onset, progression and impact of symptoms on quality of life. A specific inquiry should be made about family history, ethnicity and consanguinity (see Chapter 2 for children and Chapter 3 for adults for further details). A general medical history should also be obtained (e.g., visual symptoms, diabetes, renal and cardiovascular disease that may indicate a syndromic cause) and, importantly, exposure to aminoglycosides should be noted. In pediatric cases, it is important to cover the pregnancy and perinatal period and to exclude other important causes of hearing loss by inquiring specifically about toxoplasmosis, rubella, cytomegalovirus (CMV), herpes and syphilis infections. General developmental progression should be charted against milestones as delays may indicate global or specific language or motor delays, indicative of syndromic mitochondrial disease.

In pediatric cases, clinical examination should include inspection of the craniofacial region, including physical measurements, to exclude a number of non-mitochondrial

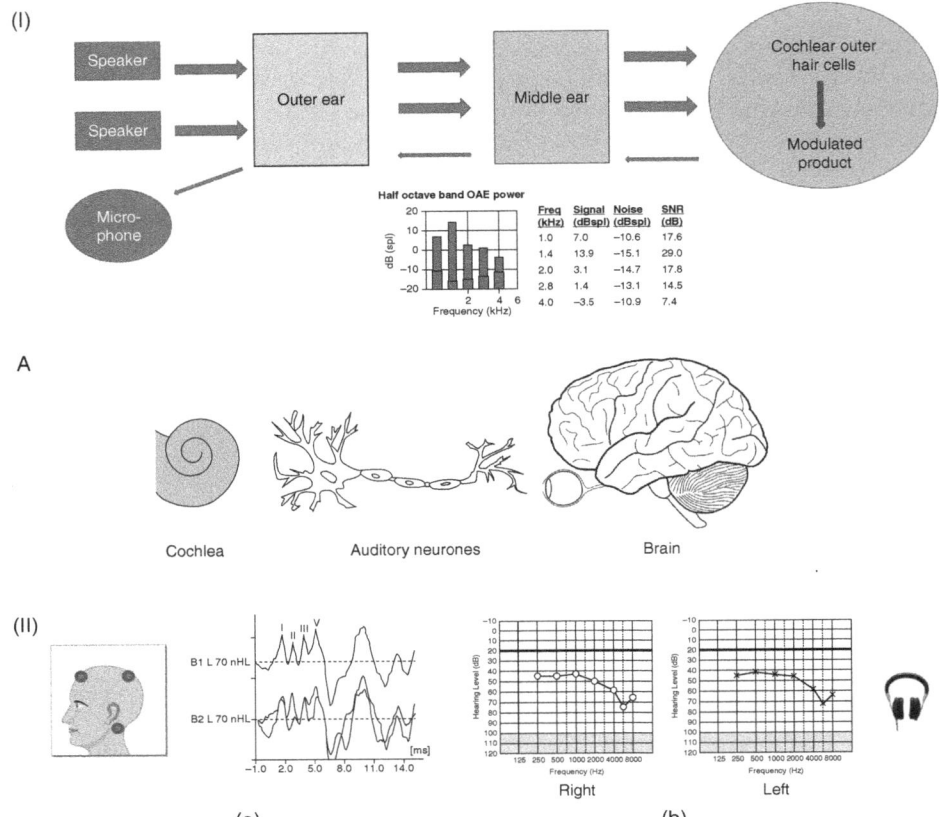

Figure 10.1 (A) Schematic diagram of the auditory conduction pathway from the cochlea via neuronal auditory brain structures to the auditory cortex
(I) Otoacoustic emissions (OAEs): A speaker-generated tone is delivered to the ear by an indwelling ear probe that also measures the modulated product (in the case of Distortion Product OAEs) generated by the stimulated cochlear outer hair cells. Cochlea function is specifically measured (represented by blue diagrammatic brackets in Panel A). (IIa) Auditory brainstem response (ABR): position of skull electrodes and representation of electrophysiological response of auditory pathway (waves I-V marked). (IIb) Pure tone audiometry (PTA): Audiograms of the right and left ears showing pan-frequency hearing loss with raised thresholds at high frequencies (lower limit of normal hearing marked with horizontal blue line). Subject responds to different frequency tones presented by headphones. (II) Both ABR and PTA measure response of the entire auditory pathway (represented by red diagrammatic brackets in Panel A)

hearing loss syndromes that are associated with defects in craniofacial development. A CT scan of the petrous temporal bones may be performed to rule out congenital abnormalities of the cochlea, particularly in cases where cochlear implantation is being considered.

Examination of the peripheral and central auditory sensory axis from cochlea to brain can be undertaken using both behavioral tests that require age-specific modifications, and physiologic tests that can be undertaken at any age (see Figure 10.1). Combining information from these tests can both quantify the degree of hearing loss and differentiate between sensory and neural components.

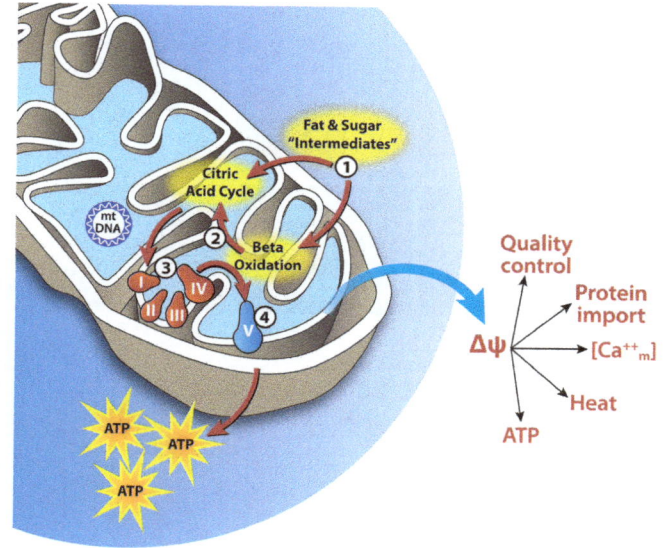

Figure 1.1 Main metabolic pathways within mitochondria.

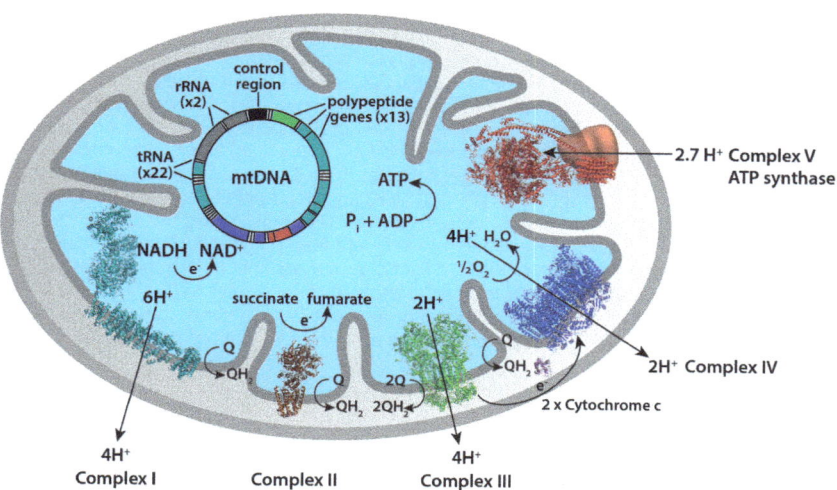

Figure 1.2 Overview of the OxPhos system (complexes I–V) and of the organization of mitochondrial DNA.

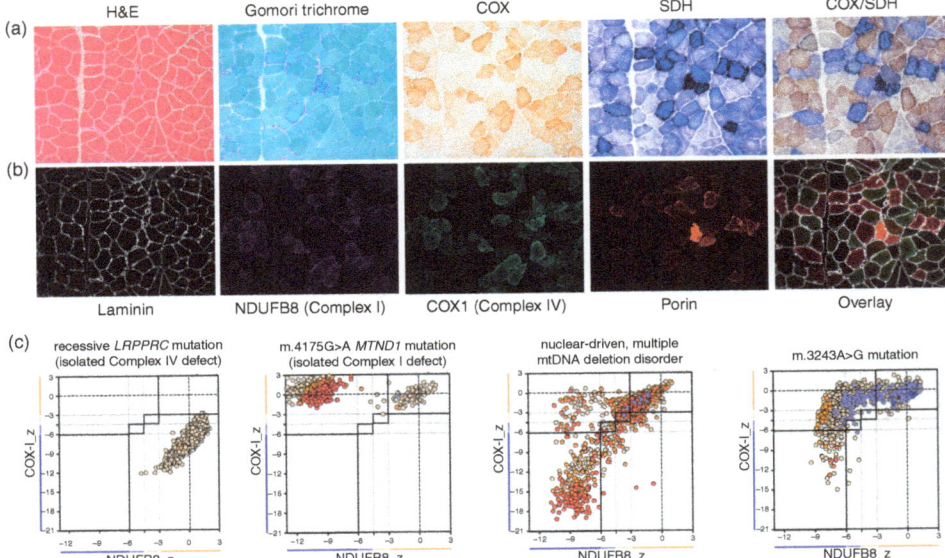

Figure 4.1 Histological, histochemical and immunohistochemical hallmarks of mitochondrial pathology in mito-chondrial disease. **A**, serial skeletal muscle (quadriceps) sections from a patient with a nuclear-driven, multiple mtDNA deletion disorder were stained for hematoxylin and eosin (H&E) and modified Gomori trichrome to assess basic muscle morphology and demonstrate the presence of ragged red fibers – a sign of abnormal mitochondrial accumulation – respectively. The individual COX, SDH and sequential COX/SDH histochemical reactions show fibers manifesting mitochondrial accumulation and focal COX-deficiency. **B**, serial sections of the same muscle biopsy sample subjected to a quadruple immunofluorescence assay that can quantitate the expression of complex I (NDUFB8 subunit), complex IV (COXI subunit), laminin (to mark muscle fiber boundaries) and a mitochondrial mass marker (porin), all within a single 10μm section (all images taken at 10X). **C**, mitochondrial respiratory chain expression profile plots (see Rocha et al. [17]) showing COX-I, NDUFB8 and porin protein levels in single muscle fibers from patients with different mitochondrial genetic defects (as indicated), highlighting the different profiles relating to mitochondrial genotype. Each dot represents the measurement from an individual muscle fiber, color-coded according to its mitochondrial mass (very low: blue; low: light blue; normal: light orange; high: orange; very high: red). Thin black dashed lines indicate the SD limits for the classification of fibers, lines next to the x and y axes indicate the levels of NDUFB8 and COX-I, respectively (beige: normal; light beige: intermediate (+); light blue: intermediate (-); blue: deficient). Bold dashed lines indicate the mean expression level of normal fibers.

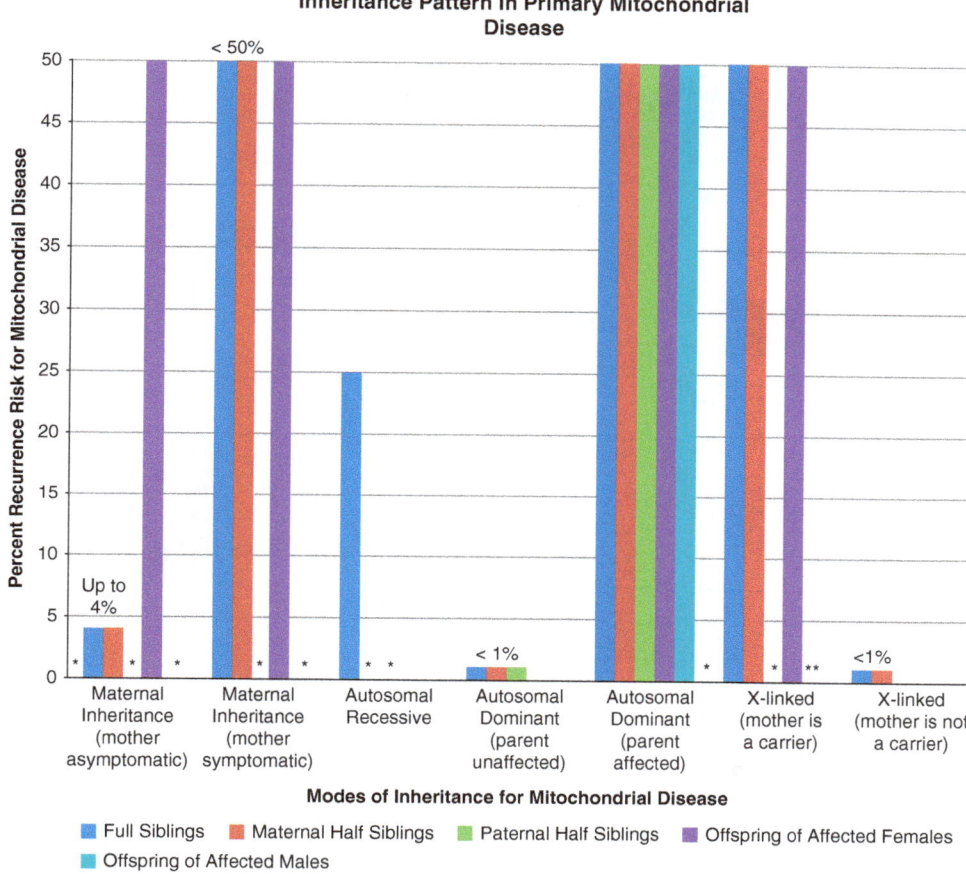

Figure 5.1 * indicates no recurrence risk.

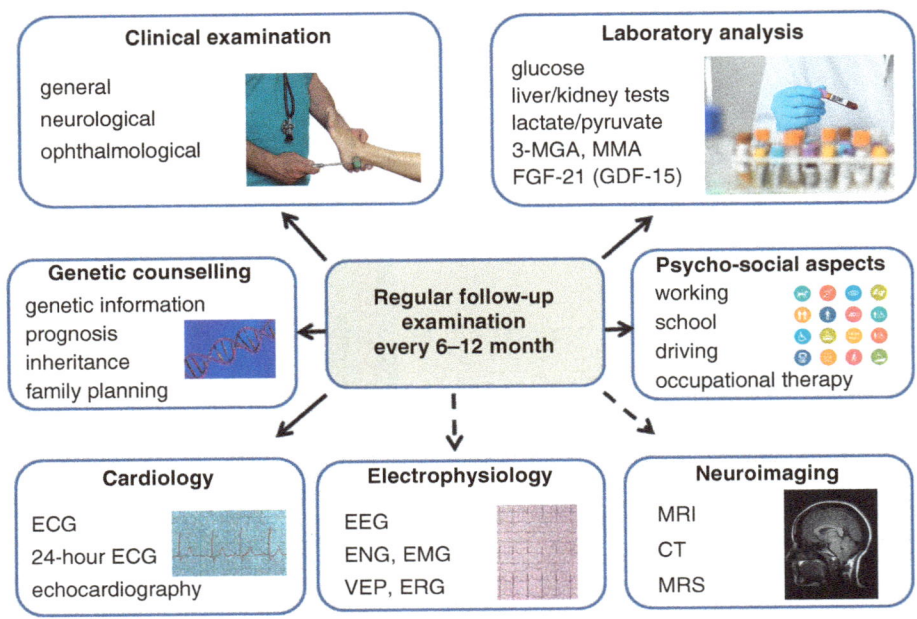

Figure 6.1 Regular follow-up of patients with mitochondrial disease.

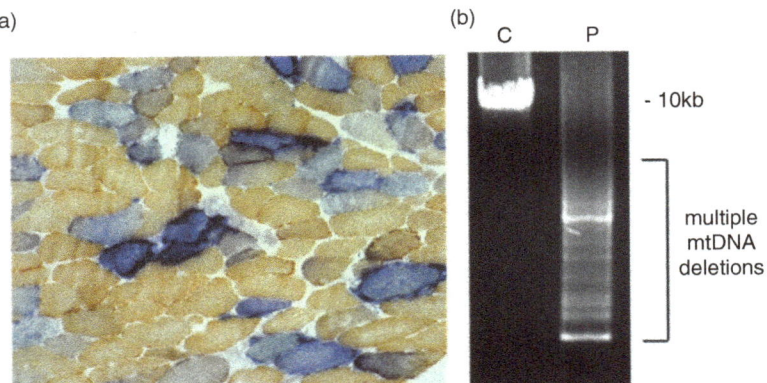

Case Report Figure 7.4 (A) Sequential COX/SDH histochemistry demonstrates a mosaic distribution of COX-deficient muscle fibers (blue) among fibers exhibiting normal COX activity (brown), with significant evidence of mitochondrial proliferation as shown by enhanced SDH reactivity around the subsarcolemmal region of the muscle fiber (ragged blue fibers). (B) Long-range PCR amplification of muscle DNA across the major arc shows significant evidence of multiple mtDNA deletions. C = Control; P = patient (adapted from *Brain* 2014 May;137(Pt. 5):1323–1336).

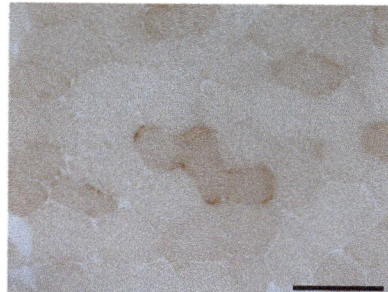

Figure 8.1 Muscle biopsy specimen shows decreased COX staining in the majority of fibers. Scale bars represent 100 μm.

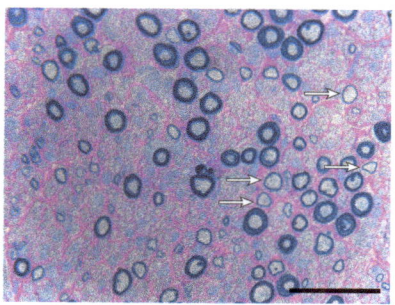

Figure 8.2 Sural nerve biopsy specimen. High-magnification view of one fascicle is shown (semithin section of resin stained with methylene blue-azure basic fuchsin). Multiple large-diameter axons with inappropriately thin myelin sheaths, suggesting demyelination, are shown by arrows. Scale bars represent 100 μm.

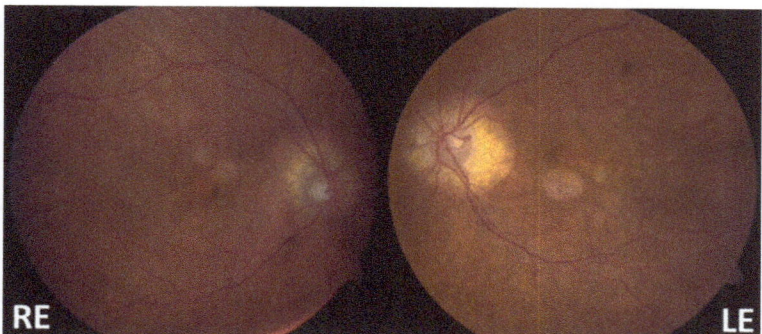

Figure 9.1 Pigmentary retinopathy in mitochondrial disease
There is generalized retinal pigment epithelial disturbance with areas of hypo- and hyper-pigmentation in this patient with Kearns-Sayre syndrome secondary to a single large-scale mtDNA deletion. A ring of peripapillary atrophy can also be observed, which is more prominent around the left optic disc. LE, left eye; RE, right eye.

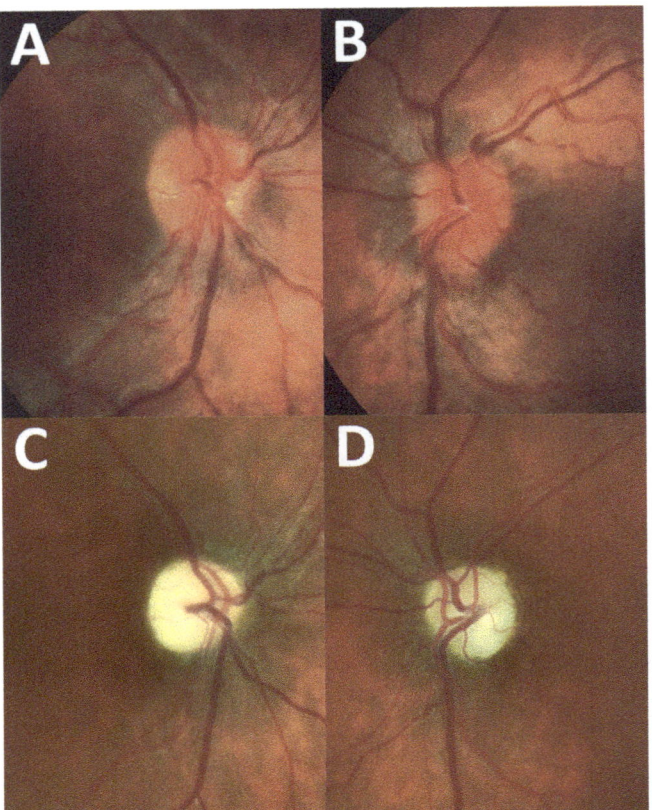

Figure 9.2 Optic disc appearance in the acute and chronic stages of LHON.
(A and B) In the acute stage, the patient's optic discs were hyperemic with peripapillary telangiectatic vessels and tortuosity of the central retinal vessels (March 2011). Mild temporal disc pallor was already evident in the right eye, and this was associated with retinal nerve fibre layer thinning within the papillomacular bundle on OCT. (C and D) The swelling of the retinal nerve fibre layer had resolved and both optic discs showed marked pallor of the neuroretinal rim (December 2013).

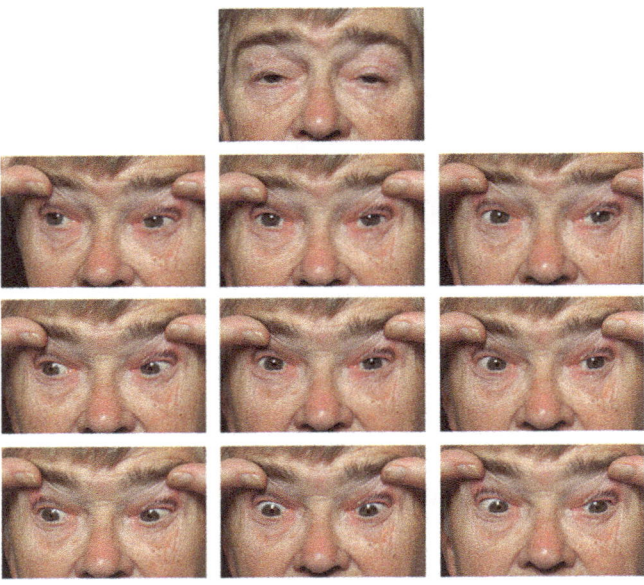

Figure 9.3 Chronic progressive external ophthamoplegia
The patient had significant bilateral ptosis and the upper eyelids were lifted to more clearly demonstrate the range of eye moments in the nine cardinal positions of gaze. There was symmetrical limitation of eye movements, which was worse on upgaze.

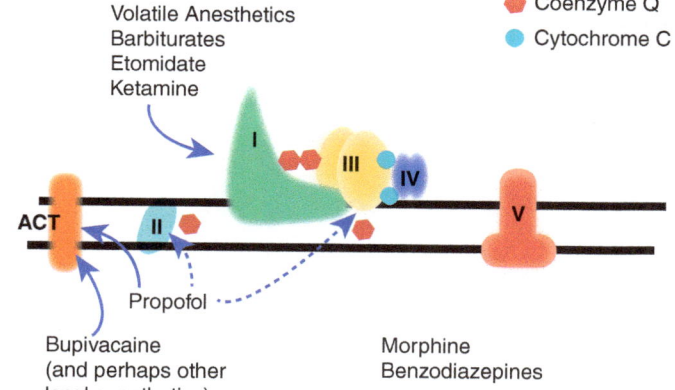

Volatile Anesthetics
Barbiturates
Etomidate
Ketamine

⬡ Coenzyme Q
🔵 Cytochrome C

ACT

I

III

IV

II

V

Propofol

Bupivacaine
(and perhaps other
local anesthetics)

Morphine
Benzodiazepines

Figure 12.2 The primary sites at which common anesthetic agents affect mitochondrial function. The two most notable agents are a) the volatile anesthetics, which are strong inhibitors of complex I function; and b) propofol, which is reported to affect multiple sites. However, the primary site of propofol action in mitochondria is felt to be inhibition of the transport of fatty acids into the matrix. Morphine and benzodiazepines are without arrows since their effects are inconsistent in different studies.

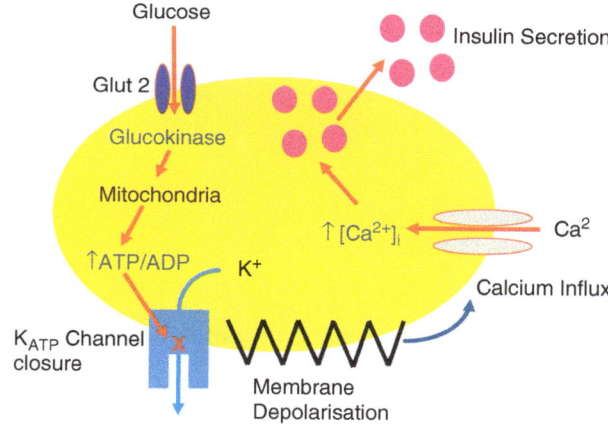

Insulin secretion from the beta-cell

Glucose

Glut 2

Glucokinase

Mitochondria

↑ATP/ADP

K+

K_{ATP} Channel
closure

Membrane
Depolarisation

↑ $[Ca^{2+}]_i$

Ca^2

Insulin Secretion

Calcium Influx

Figure 13.1 Schematic diagram depicting the key steps in glucose-stimulated insulin secretion from the pancreatic beta-cell. Glucose enters the beta-cell via the GLUT 2 transporter, and is then metabolized by the mitochondria to generate ATP. An increase in the ATP/ADP ratio leads to closure of the K_{ATP} channel, cell membrane depolarization and opening of calcium channels. Influx of calcium into the beta-cell stimulates insulin release.

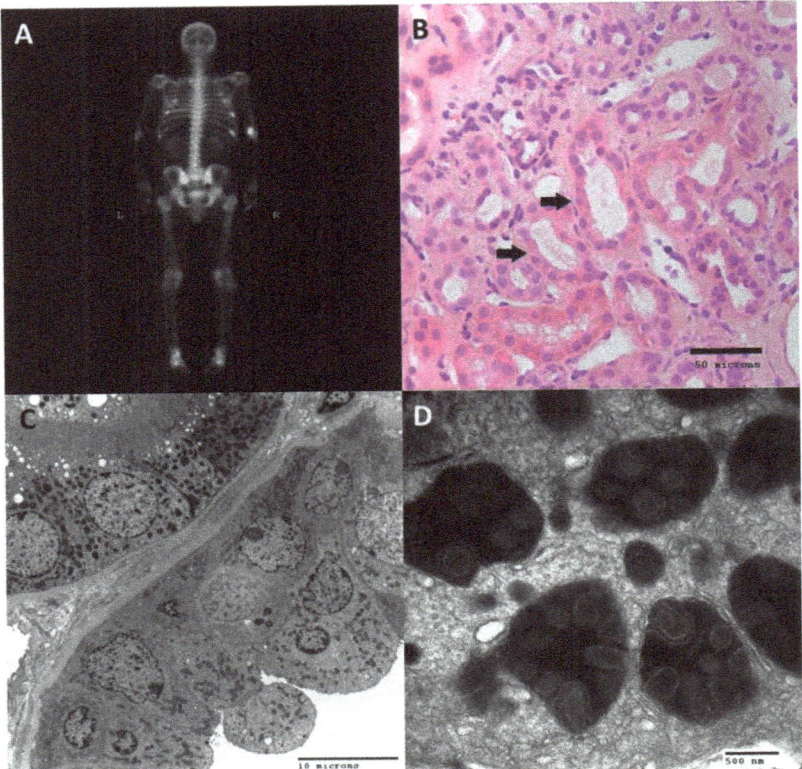

Figure 15.1 (**A**) bone scan showing evidence of osteomalacia in adult patient with tenofovir-induced Fanconi syndrome; (**B**) light micrograph of renal biopsy showing acute tubular injury with epithelial cell flattening and dedifferentiation (arrowheads); (**C** and **D**) electron micrograph of renal biopsy showing abnormal mitochondria, with a highly electron-dense, condensed matrix, and occasional loss or distortion of cristae, in the epithelial cells of proximal tubules.

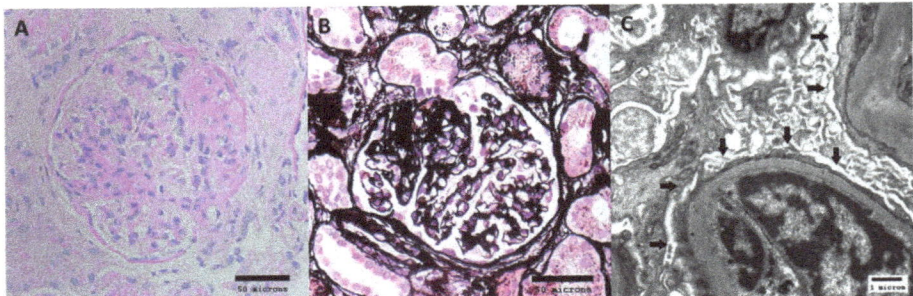

Figure 15.2 (**A** and **B**) Renal biopsy showing segmental sclerosis of a glomerulus by light microscopy (hematoxylin and eosin stain, **A,** and silver stain, **B**); this lesion was found to affect 5 of 40 glomeruli samples, consistent with a diagnosis of focal segmental glomerulosclerosis (FSGS); (**C**) electron microscopy of glomerular epithelial cells (podocytes) showing extensive foot process effacement (arrowheads), which results in malfunction of the capillary filtration barrier in the glomerulus and the clinical finding of proteinuria.

Behavioral Tests

Pure tone audiometry (PTA): the cornerstone of diagnostic audiology, is a subjective behavioral test of a patient's ability to detect a pure tone signal presented over a range of frequencies (typically 0.25–8 KHz), examining the function of the whole auditory axis from cochlea to brain. Hearing thresholds are presented as an audiogram, a graphical representation of threshold intensity as a function of tone frequency. PTA tests both air and bone conduction, enabling the differentiation between sensorineural and conductive hearing losses (bone conduction bypasses the outer and middle ears and hence mitigates the effect of conductive losses). Typically, mitochondrial hearing loss produces isolated high-frequency loss (down-sloping audiogram) or in more severe cases, a pan-frequency loss with no difference between the air and bone thresholds typical of a sensorineural loss (see Case Reports 10.1 and 10.2).

In pediatric cases, age-appropriate behavioral hearing tests should be used as follows:

<u>0–6 months</u>: *Behavioral observation audiometry*: uses response to intense stimuli, but can only rule out profound hearing losses.

<u>6–18 months</u>: *Distraction testing*: Uses two testers, a distractor and a presenter, and allows testing of a range of sounds.

<u>6–30 months</u>: *Visual reinforcement audiology*: is a reward-conditioned response to visual stimulation and can be used to examine specific speech frequencies.

<u>30–36+ months</u>: *Performance audiometry/play audiometry*: uses a conditioned response (e.g., picking up a toy) to a range of warble tones.

<u>5–7+ years:</u> Traditional pure tone audiometry can be used in cooperative children.

Physiologic Tests

Tympanometry: is an examination of pressure transmission of the middle ear. A probe is used to deliver a tone of 226 Hz to the tympanic membrane while an air pump changes the air pressure range in the ear canal. The proportion of sound energy reflected as the pressure changes is used to infer middle ear compliance, which can be used to differentiate between sensorineural and conductive hearing loss.

Otoacoustic emissions (OAEs): are cochlea-generated sound waves resulting from the mechanical frequency selectivity of the outer hair cells. These can be either spontaneous or evoked (by a signal frequency applied by an earphone) and are measured using a flexible indwelling ear probe. In clinical practice evoked emissions (either transient, TEOAEs, or distortion product, DPOAEs) are routinely used to determine peripheral (cochlear) auditory function. The presence of OAEs confirms that the cochlear sensitivity in the relevant region is 20–40 dB HL or better (normal hearing thresholds are classified as between −10 and +20 dB HL).

Auditory brainstem response (ABR): is an evoked electrophysiological response generated by an auditory stimulus, usually a tone pip or broadband click, which can be used to measure the resulting neural activity of the cochlea-brain axis. The amplitude of the response is measured through surface electrodes placed at the vertex and ear lobe. Responses are then plotted against time, giving seven morphologically characteristic waveforms (I–VII) representing sequential stimulation of auditory brain structures. The resulting waveforms can be examined for amplitude, interpeak latency, absolute latency and interaural latency difference. The diagnosis of an auditory neuropathy depends on increased interpeak latencies in the presence of OAEs with increased thresholds on pure tone audiometry. ABR may not be technically possible in patients with severe sensorineural hearing loss.

Laboratory Investigations

Mitochondrial mutation analysis is primarily performed on DNA extracted from blood and skeletal muscle. It is also possible to perform analysis on hair follicles, buccal mucosa and urinary sediment. Specific genetic tests should be conducted where appropriate to examine a number of well-characterized variants and mitochondrial DNA rearrangements associated with both mitochondrial associated non-syndromic and syndromic hearing loss (discussed later in this chapter).

Genetic Etiology of Mitochondrial-Associated Hearing Loss

Mitochondrial disease results from mutation in nuclear genes encoding mitochondrial proteins or from mutation of mitochondrial DNA. Subsequently, mitochondrial hearing loss can be inherited following either a Mendelian or maternal inheritance pattern.

Mitochondrial DNA mutations associated with non-syndromic hearing loss: Mutations causing non-syndromic hearing loss are found in genes encoding components of mitochondrial translation such as *MTRNR1* that encodes the mitochondrial 12S ribosomal RNA and the *MTTS1* gene encoding the tRNA for Ser$^{(UCN)}$.

Variants in the mitochondrial *MTRNR1* gene coding for the 12S subunit of the ribosomal RNA have been linked to both maternally transmitted non-syndromic and aminoglycoside associated deafness [14]. These variants include m.1555A>G, m.1095 T>C, m.1494C>T and variants at position 961. Studies have shown the m.1555A>G and m.1494C>T cause a local conformational change in the aminoacyl tRNA binding site (A site) of 12S rRNA that facilitates increased aminoglycoside binding to the ribosome. This subsequently affects the efficiency of mitochondrial translation by modulation of codon–anticodon interactions [15]. The initiation of apoptosis by (ROS) oxygen species activated caspases may partially explain the link between mitochondrial mistranslation and subsequent cochlea hair cell death [16].

Hearing loss in carriers of m.1555A>G is usually the sole presenting phenotype and is typically bilateral, with sensorineural loss at high frequencies without vestibular dysfunction. Hearing loss in carriers after aminoglycoside administration has traditionally been believed to be 100 percent penetrant with more profound loss occurring in patients receiving aminoglycosides when less than 10 years of age. Recently, however, there have been accounts of carriers receiving multiple doses of aminoglycosides with unaffected hearing levels [17]. The penetrance and age of onset of hearing loss for those who have not received aminoglycosides is variable and may be influenced by other environmental factors, e.g., noise exposure or nuclear modifier genes (as discussed later in this chapter).

Several mutations in the *MTTS1* gene encoding tRNA Ser$^{(UCN)}$, including m.7445A>G, 7472insC, m.7510 T>C and m.7445 G>, have been associated with non-syndromic mitochondrial hearing loss. These variants are proposed to lead to a reduction in tRNA processing and a decline in cellular oxidative phosphorylation [18]. Variants in *MTTS1* have also been associated with syndromic cases, for example, the m.7445A>G variant causes hearing loss combined with palmoplantar keratoderma, while m.7512A>G has also been associated with forms of MERRF (see later in this chapter) [19, 20].

Mitochondrial DNA mutations associated with syndromic hearing loss: The commonest forms of mitochondrial syndromic hearing loss are associated with the complex neuromuscular syndromes Kearns-Sayre syndrome, mitochondrial encephalomyopathy, lactic acidosis and stroke-like episodes (MELAS) and myoclonic epilepsy with ragged red fibers (MERRF) that are caused by large mitochondrial DNA rearrangements and variants in mitochondrial tRNA genes including *MT-TL1* (tRNA$^{Leu(UUR)}$), *MT-TK* (tRNA^{Lys}) and *MT-TE* (tRNA^{Glu}), respectively. The common MELAS variant, m.3243A>G, is also associated with maternally inherited diabetes and deafness (MIDD) consisting of a triad of maternally inherited diabetes with a normal body mass index, hearing impairment and retinal dystrophy. MIDD can also be associated with mutations in the *MT-TK* and *MT-TE* genes.

Nuclear genes associated with mitochondrial hearing loss: Given that the majority of mitochondrial proteins is encoded by the nuclear genome, it follows that nuclear-encoded mitochondrial proteins may act to modify the phenotype of primary mitochondrial DNA mutations or cause hearing loss directly. *TFB1M, MTO1, GTPBP3* and *TRMU* have all been suggested as possible candidates for nuclear-modifying factors of the m.1555A>G variant. These variants may act synergistically to propagate mitochondrial dysfunction by exacerbating the effect of mitochondrial DNA mutations through their role in modification of mitochondrial rRNA and tRNAs (reviewed in [21]). A number of nuclear genes cause hearing loss directly as a feature of syndromic mitochondrial disease by dysregulating mitochondrial DNA replication or repair (mitochondrial DNA maintenance). *OPA1* encodes a dynamin-related GTPase that regulates both mitochondrial fusion and mitochondrial DNA maintenance. Mutations in *OPA1* are commonly associated with dominant optic atrophy (DOA), a selective degeneration of retinal ganglion cells; however, a subset of missense mutations causes a DOA-plus phenotype presenting with optic atrophy in conjunction with myopathic features, progressive external ophthalmoplegia (PEO) and hearing impairment. Interestingly, it has been shown that different mutations within the gene result in hearing loss through divergent mechanisms with haploinsufficiency mutations primarily causing a cochlear loss whereas missense mutations cause primarily an auditory neuropathy [10]. *POLG* encodes the unique replicative mitochondrial DNA polymerase and mutations in this gene cause hearing loss in the context of complex neurological phenotypes including PEO, Alpers disease and sensory ataxia with neuropathy, dysarthria and ophthalmoplegia (SANDO). Other mtDNA maintenance genes associated with mitochondrial deafness include *MPV17, SUCLA2, RRM2B* and *C10ORF2*. Similarly, mutations in genes involved in apoptosis (*SMAC/Diablo*) and OXPHOS complex assembly (*BCS1L, COX10*) have also been implicated.

Acquired hearing loss: Mitochondria are an important source of cellular ROS generated as a by-product of mitochondrial oxidative phosphorylation. Noise exposure can drive increased mitochondrial ROS production that may exceed the buffering capacity of cellular antioxidants, leading to oxidative stress. Subsequent cochlea hair cell death is then driven in part by mitochondrial-mediated activation of cellular apoptosis. There is some evidence that treatment with antioxidants can attenuate hair cell death and subsequent threshold shifts.

In addition, mitochondria play a fundamental role in cellular homeostasis and hence cellular aging. Due to the vulnerability of the mitochondrial genome to mutation, genetic changes are known to accumulate throughout life. However, dissecting which of these genetic changes is specific to age-induced hearing loss has proved challenging. A specific common deletion (mtDNA 4977bp) has been seen to be overrepresented in temporal bone material from patients with age-onset hearing loss as compared to aged matched controls [4]. This has paralleled work in several mouse models that suggest accumulation of mitochondrial DNA mutations and increased oxidative stress in the cochlea contribute to age-related hearing loss (ARHL). However a recent larger analysis of the mitochondrial genome in 400 individuals (200 with normal hearing and 200 with poor hearing) found no association between mitochondrial DNA mutation load and ARHL [22].

Clinical Management

There is currently no medical treatment for sensorineural hearing loss. Amplification by hearing aids used with or without hearing assistive technology systems (e.g., infrared/FM systems, induction loop systems) is the mainstay of treatment. In cases of severe hearing loss, cochlear implantation may be indicated and has been used effectively in patients with mt.3243A>G, *OPA1*, Kearns-Sayre syndrome and mitochondrial neurogastrointestinal encephalomyopathy (MNGIE) syndrome [10 23–25]. Individuals known to harbor the m.1555A>G mutation should avoid aminoglycoside exposure.

CASE REPORTS

Case Report 10.1

Child A is a six-year-old Caucasian female with a diagnosis of cystic fibrosis. She was born at term with a low birth weight (2.3 kg) although without perinatal problems. She has one healthy younger sibling and her family history is otherwise unremarkable for systemic disease. Prior to diagnosis there had been no concern with hearing or language delay and she had passed newborn hearing screening. Child A's parents had normal hearing and her paternal grandfather had mild hearing loss suggestive of age-onset hearing loss. Since infancy she had suffered recurrent respiratory tract infections requiring multiple admissions to hospital. As part of treatment for pneumonia, she received two courses of intravenous amikacin (10 mg/kg bodyweight/14 days per course). One year after diagnosis, she was noted to have speech regression and very limited response to auditory stimuli. The results of her PTAs are shown in Figure 10.2.

Self-Assessment Questions

- List three possible causes of Child A's deafness.
- List three investigations to document Child A's hearing levels.
- What are these likely to show?
- Suggest one possible candidate gene that could be sequenced.
- How could Child A be managed?

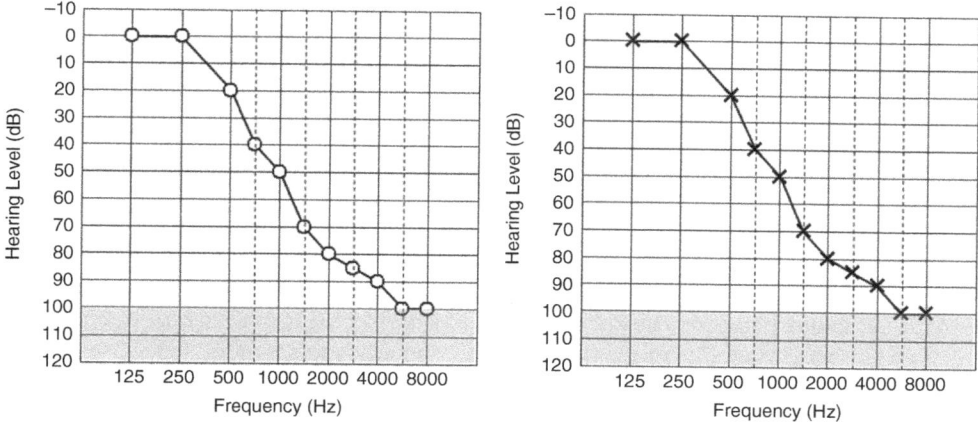

Answers

Child A's deafness could be caused by:

- m.1555A>G mutation in association with aminoglycoside administration.
- Other genetic non-mitochondrial causes of delayed-onset hearing loss, for example, connexin 26 (*GJB2*) mutation (in some cases) or *SLC26A4* (associated with enlarged vestibular aqueduct).
- Direct cochleotoxicity of aminoglycosides without obvious underlying genetic susceptibility.

Full audiological assessment should be undertaken with:

- Standard pure tone audiometry (PTA) (0.25–8 KHz). This enables the level of hearing loss to be determined across these frequencies. The character of the audiogram may be suggestive of the underlying pathology, e.g., sensorineural or conductive (in this case, the absence of a difference in air and bone conduction indicates a sensorineural loss). In general, children over the age of five years are cooperative enough to perform the test.
- Extended high-frequency audiometry (9–16 KHz) permits investigation of higher frequencies that are affected more severely by aminoglycoside cochleotoxicity.
- Tympanometry in combination with the results of the PTA is used to rule out conductive causes of hearing loss.
- Distortion product otoacoustic emissions (DPOAEs) (0.8–8 KHz) will specifically examine the cochlea function (outer hair cell function).
- Auditory brainstem electrophysiology can determine whether there are underlying neuropathic or central causes for deafness.

Child A's audiogram shows a bilateral moderate-severe, mid-frequency sloping to severe-profound, high-frequency sensorineural hearing loss. Tympanometry reveals normal middle ear compliance, confirming sensorineural rather than conductive loss. DPOAEs are absent bilaterally typical of cochlea hearing loss. ABRs are unobtainable due to the severity of the hearing loss (auditory brain structures are not adequately stimulated to the detection level by the reduced function of the cochlea).

Management comprises audiological rehabilitation using a family-centered interdisciplinary approach. The mainstays of treatment are:

- Hearing aids. Bilateral behind-the-ear aids provide optimal amplification.
- Bilateral cochlear implantation should also be considered. Current guidelines require a bilateral sensorineural hearing loss of more than 90 dB guidelines at 2–4 KHz and after failing a hearing aid trial of three months. However, in cases of severe loss not meeting the guideline, as in the case of Child A, individual factors such as patient quality of life are also taken into consideration.

Child A was entered into the services of the cochlear implant team. A high-resolution CT scan of both temporal bones was performed. This confirmed the absence of any malformations of the auditory tract or abnormality preventing correct siting of the electrode array, and implantation surgery was performed. Although complications are rare, these include skin infection, worsening tinnitus, vestibular disequilibrium and transitory facial nerve weakness. Device activation was undertaken four weeks after implantation. Following this Child A was entered into a one-year structured program of intensive training and rehabilitation.

Case Report 10.2

Adult A is a 34-year-old with a background of congenital short stature, insulin-dependent diabetes, cataracts, leg weakness and fatigue. She has no history of encephalopathy or strokes. She reports gradually deteriorating hearing since the age of 20 years. Currently she is finding no benefit in the use of hearing aids. Her mother was also diabetic and had less severe deafness. She has one sister who had severe-profound deafness from infancy as well as severe mental retardation.

CT petrous temporal bones revealed no abnormalities of the inner ear. Figure 10.3 shows the result of Adult A's pure tone audiogram.

Self-Assessment Questions

- Describe the result of Adult A's audiogram.
- How could the cause of this hearing loss be more thoroughly investigated?

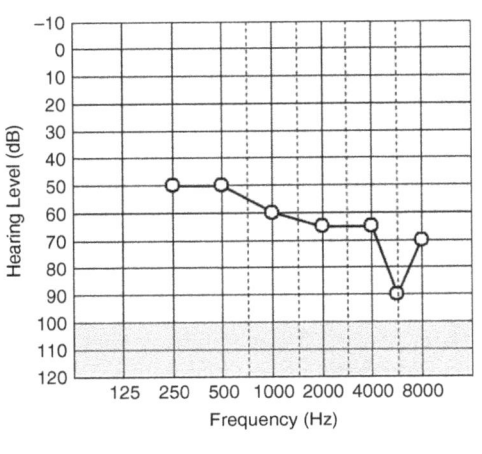
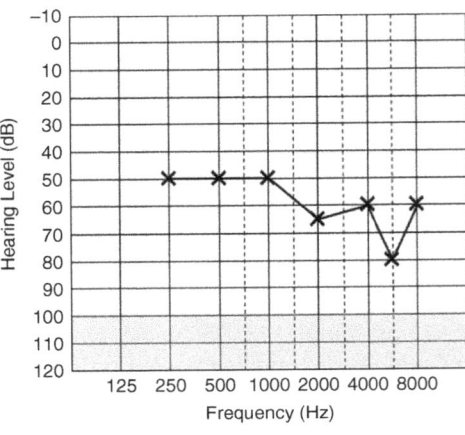

- Suggest a possible diagnosis for Adult A's hearing loss.
- How could her hearing loss be managed?

Answers: The audiogram shows a bilateral, moderate, mid-frequency sloping to moderate-severe sensorineural hearing loss.

Given the systemic features and apparent maternal transmission of disease, a full investigation of mitochondrial disease should be undertaken. This should initially include analysis of mtDNA from the patient's blood, and possibly a subsequent muscle biopsy for mtDNA mutation analysis and respiratory enzyme activity.

Adult A's hearing loss occurs in the context of systemic disease with apparent matrilineal segregation. Laboratory studies revealed a heteroplasmic m.3243A>G mutation in blood (45 percent heteroplasmy) confirming the diagnosis of mitochondrial deafness.

Provision of adequate sound amplification by appropriately fitting hearing aids is the mainstay in management of moderate-severe hearing loss in an adult. Additional hearing-assistive technology includes hearing loops, visual trigger aids and specialist alarm clocks and telephones. There are also a number of hearing support groups that patients should be encouraged to attend.

References

1. UK AAoHL. wwwactiononhearinglossor guk/your-hearing/about-deafness-and-hear ing-loss/statisticsaspx

2. Jacobs HT, Hutchin TP, Kappi T, et al. Mitochondrial DNA mutations in patients with postlingual, nonsyndromic hearing impairment. *Eur J Hum Genet* 2005;**13** (1):26–33. doi: 10.1038/sj.ejhg.5201250

3. Morton CC, Nance WE. Newborn hearing screening – a silent revolution. *The New England Journal of Medicine* 2006;**354** (20):2151–2164. doi: 10.1056/ NEJMra050700

4. Bai U, Seidman MD, Hinojosa R, et al. Mitochondrial DNA deletions associated with aging and possibly presbycusis: A human archival temporal bone study. *The American Journal of Otology* 1997;**18** (4):449–453.

5. Chinnery PF, Elliott C, Green GR, et al. The spectrum of hearing loss due to mitochondrial DNA defects. *Brain: A Journal of Neurology* 2000;**123** (Pt. 1):82–92.

6. Sue CM, Lipsett LJ, Crimmins DS, et al. Cochlear origin of hearing loss in MELAS syndrome. *Annals of Neurology* 1998;**43** (3):350–359. doi: 10.1002/ana.410430313

7. Liu Y, Xue J, Zhao D, et al. Audiological evaluation in Chinese patients with

8. Kullar PJ, Quail J, Lindsey P, et al. Both mitochondrial DNA and mitonuclear gene mutations cause hearing loss through cochlear dysfunction. *Brain: A Journal of Neurology* 2016;**139**(Pt 6):e33. doi: 10.1093/brain/aww051

9. Ceranic B, Luxon LM. Progressive auditory neuropathy in patients with Leber's hereditary optic neuropathy. *J Neurol Neurosurg Psychiatry* 2004;**75** (4):626–630.

10. Santarelli R, Rossi R, Scimemi P, et al. OPA1-related auditory neuropathy: Site of lesion and outcome of cochlear implantation. *Brain: A Journal of Neurology* 2015;**138**(Pt 3):563–576. doi: 10.1093/ brain/awu378

11. Edmonds JL, Kirse DJ, Kearns D, et al. The otolaryngological manifestations of mitochondrial disease and the risk of neurodegeneration with infection. *Arch Otolaryngol Head Neck Surg* 2002;**128** (4):355–362.

12. Chen JC, Tsai TC, Liu CS, et al. Acute hearing loss in a patient with mitochondrial myopathy, encephalopathy, lactic acidosis and stroke-like episodes (MELAS). *Acta neurologica Taiwanica* 2007;**16**(3):168–172.

mitochondrial encephalomyopathies. *Chin Med J (Engl)* 2014;**127**(12):2304–2309.

13. Uimonen S, Moilanen JS, Sorri M, et al. Hearing impairment in patients with 3243A–>G mtDNA mutation: Phenotype and rate of progression. *Hum Genet* 2001;**108**(4):284–9.

14. Estivill X, Govea N, Barcelo E, et al. Familial progressive sensorineural deafness is mainly due to the mtDNA A1555 G mutation and is enhanced by treatment of aminoglycosides. *American Journal of Human Genetics* 1998;**62**(1):27–35.

15. Hobbie SN, Bruell CM, Akshay S, et al. Mitochondrial deafness alleles confer misreading of the genetic code. *Proc Natl Acad Sci U S A* 2008;**105**(9):3244–3249. doi: 10.1073/pnas.0707265105

16. Priuska EM, Schacht J. Formation of free radicals by gentamicin and iron and evidence for an iron/gentamicin complex. *Biochem Pharmacol* 1995;**50**(11):1749–1752.

17. Al-Malky G, Suri R, Sirimanna T, et al. Normal hearing in a child with the m.1555A>G mutation despite repeated exposure to aminoglycosides. Has the penetrance of this pharmacogenetic interaction been overestimated? *International Journal of Pediatric Otorhinolaryngology* 2014;**78**(6):969–973. doi: 10.1016/j.ijporl.2014.02.015

18. Guan MX, Enriquez JA, Fischel-Ghodsian N, et al. The deafness-associated mitochondrial DNA mutation at position 7445, which affects tRNASer(UCN) precursor processing, has long-range effects on NADH dehydrogenase subunit ND6 gene expression. *Mol Cell Biol* 1998;**18**(10):5868–5879.

19. Sevior KB, Hatamochi A, Stewart IA, et al. Mitochondrial A7445 G mutation in two pedigrees with palmoplantar keratoderma and deafness. *Am J Med Genet* 1998;**75**(2):179–185.

20. Nakamura M, Nakano S, Goto Y, et al. A novel point mutation in the mitochondrial tRNA(Ser(UCN)) gene detected in a family with MERRF/MELAS overlap syndrome. *Biochemical and Biophysical Research Communications* 1995;**214**(1):86–93. doi: 10.1006/bbrc.1995.2260

21. Luo LF, Hou CC, Yang WX. Nuclear factors: roles related to mitochondrial deafness. *Gene* 2013;**520**(2):79–89. doi: 10.1016/j.gene.2013.03.041

22. Bonneux S, Fransen E, Van Eyken E, et al. Inherited mitochondrial variants are not a major cause of age-related hearing impairment in the European population. *Mitochondrion* 2011;**11**(5):729–734. doi: 10.1016/j.mito.2011.05.008

23. Li JN, Han DY, Ji F, et al. Successful cochlear implantation in a patient with MNGIE syndrome. *Acta Otolaryngol* 2011;**131**(9):1012–1016. doi: 10.3109/00016489.2011.579623

24. Yamaguchi T, Himi T, Harabuchi Y, et al. Cochlear implantation in a patient with mitochondrial disease–Kearns-Sayre syndrome: A case report. *Adv Otorhinolaryngol* 1997;**52**:321–3.

25. Scarpelli M, Zappini F, Filosto M, et al. Mitochondrial sensorineural hearing loss: A retrospective study and a description of cochlear implantation in a MELAS patient. *Genet Res Int* 2012:287432. doi: 10.1155/2012/287432

Cardiovascular Medicine

Michael J. Keogh, Hannah E. Steele and Patrick F. Chinnery

Introduction

The myocardium is an exceptionally active tissue, and highly reliant on aerobic metabolism. Cardiac pathology is therefore a frequent finding in patients with mitochondrial disease, and while the proportion of patients affected varies by the underlying genetic abnormality, when taken together cardiac abnormalities may affect up to 65 percent of patients with mitochondrial disorders [1]. This equates to a population prevalence of cardiac dysfunction secondary to a mitochondrial disease of ~1/10,000 [2], therefore affecting ~6,500 individuals in the United Kingdom and ~30,000 in the United States.

While the majority of cardiac pathology in mitochondrial disease occurs in polysymptomatic individuals (i.e., patients with established symptoms in additional organ systems, e.g., neurological and ophthalmological), cardiac manifestations can present as the primary feature of disease. More commonly, however, cardiac pathology develops silently and insidiously, necessitating careful clinical surveillance and screening. Additionally, given that the severity of cardiac dysfunction may not mirror symptom severity in other organs, clinicians treating any facet of a mitochondrial disease should be aware of the importance of understanding, investigating and managing cardiac complications.

The aim of this chapter is to build knowledge of the cardiac features of mitochondrial disorders, and to act as a practical resource for all clinicians encountering patients with mitochondrial disorders.

Myocardial Disorders

Hypertrophic cardiomyopathy: Hypertrophic cardiomyopathy (HCM) is the most common form of cardiomyopathy observed across the spectrum of mitochondrial disease, present in up to 40 percent of all mitochondrial patients [3, 4]. The majority of cases of mitochondrial cardiomyopathies occur as part of a clinical spectrum of disease, usually after the suspicion of a mitochondrial disorder has been raised or a diagnosis established. The clinical presentation is broad, ranging from asymptomatic presentations where the condition is detected with cardiac surveillance (e.g., yearly echocardiograms), through to presentation with severe left ventricular dysfunction or even sudden cardiac death. When chronic symptoms are present, they are generally consistent with those observed with idiopathic cases of left ventricular (LV) impairment, with patients describing exercise limitation and dyspnea [3]. However, given the coexistence of cognitive and physical impairment due to neuromuscular disease, or breathlessness linked to exercise due to lactic acidosis, in many patients it is easy to see how symptoms such as exercise limitation could be falsely misattributed, and hence regular cardiac surveillance through echocardiograms is paramount (discussed later in this chapter and in Figure 11.1).

Homoplasmic point mutations: It is rare in our experience for a cardiomyopathy to be either the presenting or sole feature of a mitochondrial disorder. However, in a small number of cases, an isolated cardiomyopathy resulting from a homoplasmic mutation (m.4300A>G) in mt-tRNA^{Ile} (*MTTI*) has been described [5]. To date, however, no other homoplasmic mtDNA variants have been established as causing a cardiomyopathy [6], suggesting that isolated cardiac phenotypes are likely to be rare.

Heteroplasmic mtDNA point mutations: Mitochondrial point mutations causing cardiomyopathies usually occur in mitochondrial transfer RNAs (mt-tRNAs) genes such as *MTTL1* (originally described in mitochondrial encephalomyopathy with lactic acidosis) and *MTTL1* and *MTTK* (causing myoclonic epilepsy with ragged red fibers). Based on current knowledge, approximately 50 percent of patients with these mutations will eventually develop a cardiomyopathy. Although rigorous epidemiological data are not available, it does appear that mutations in non–mt-tRNA genes (e.g., mt-rRNAs and protein coding genes) are significantly less likely to result in cardiac involvement. However regular surveillance and clinical vigilance should continue in all genotypes (Figure 11.1).

Recent studies with cardiac MRI (CMR) suggest that patients with m.3243A>G show a specific morphological pattern of their cardiomyopathy, with concentric hypertrophy and patchy intramural late-gadolinium-enhancement (LGE) in the myocardium [4], differing from that seen in chronic progressive external ophthalmoplegia (CPEO). Whether these differing morphologies are true for all pathogenic mtDNA point mutations and mtDNA deletions remains unclear. Furthermore, understanding and estimating the clinical progression of mtDNA point mutation–related cardiomyopathies is difficult due to the rarity of most pathogenic mtDNA mutations, variability in tissue heteroplasmy levels and lack of longitudinal cardiac screening data in patients.

Small cohort studies suggest that around 30 percent of patients with biochemical evidence of a mitochondrial disorder (and therefore likely to be attributable to either a nuclear DNA or mtDNA defect) have evidence of left ventricular hypertrophy (LVH) at a mean age of 36 [3]. However, the true prevalence of hypertrophic cardiomyopathy may be higher than this given the recent finding of increased LV mass only when indexed to body surface area in mt.3243A>G carriers.

For patients with the m.3243 A>G mutation, a positive correlation between skeletal muscle heteroplasmy levels and left ventricular mass [7] was noted, and over a seven-year follow-up period there was a negative correlation between the degree of LVH and systolic function [8]. Taken together, these data suggest that the burden of heteroplasmy is likely to be the main determinant of the severity of cardiac sequelae following diagnosis in m.3243A>G. Whether this observation is true for all point mutations is unknown. In addition, left-ventricular outflow obstruction (LVOTO) appears to develop rarely in cardiomyopathies secondary to point mutations in mtDNA, which is in contrast to non-mitochondrial cardiomyopathies such as HCM [8].

Mitochondrial DNA deletions: Patients with Kearns-Sayre syndrome (KSS) and CPEO (usually caused by single large mtDNA deletions) are also prone to LVH, which affects around 60 percent of patients. Recent studies with CMR suggest that these patients show intramural

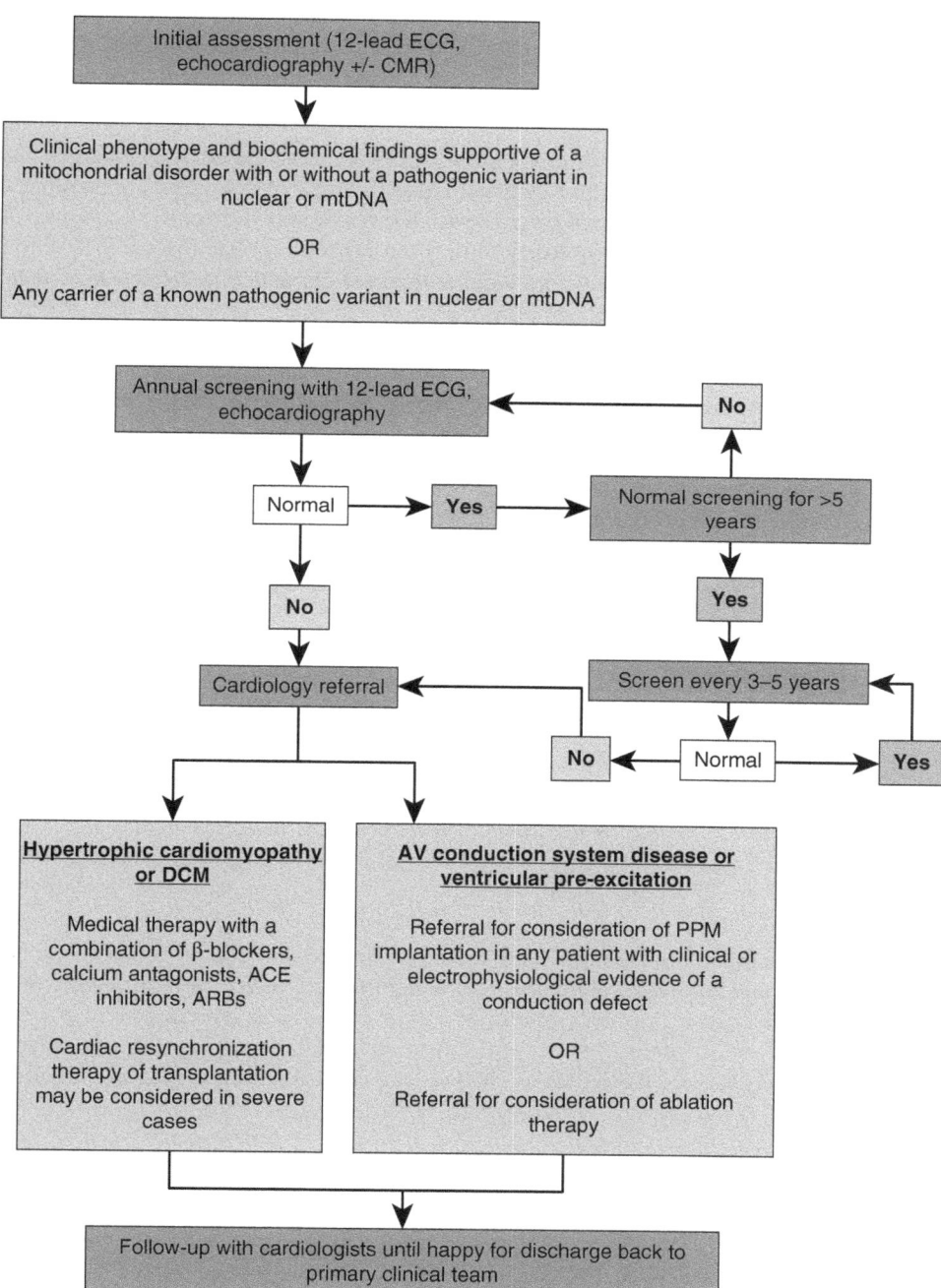

Figure 11.1 Cardiac screening and surveillance algorithm for patients with confirmed or suspected mitochondrial disorders

Blue boxes represent screening or clinical sessions. Yellow boxes indicate key moments of clinical decision-making.
Key. CMR: Cardiac MRI, DCM: Dilated cardiomyopathy, AV: Atrioventricular, PPM: Permanent pacemaker, ACE: Angiotensin-converting enzyme, ARB: Angiotensin receptor blocker

late gadolinium enhancement (LGE) in the basal and LV inferolateral wall with evidence of concentric remodeling, which is distinct from the cardiac manifestations seen in m.3243A>G carriers [4]. Data on the clinical course of cardiomyopathies in these patients are lacking.

Dilated cardiomyopathy: Dilated cardiomyopathy (DCM) is characterized by an enlarged and weakened left ventricle. It is a rare finding in patients with mtDNA disorders: for example, it is observed in fewer than 5 percent of patients with Kearns-Sayre syndrome [9], despite ~60 percent of these patients eventually developing some form of cardiac disease [1]. Where DCM has been observed in patients with mitochondrial disorders, it is likely that these individuals have developed secondary progression of a preexisting hypertrophic cardiomyopathy. However, the observation of DCM at diagnosis in some patients with high levels of cardiac heteroplasmy (90–95 percent mutant load) [10] suggests that particularly high heteroplasmy levels may favor the development of a DCM rather than LVH. This remains to be established via larger cohort studies.

Left ventricular non-compaction: Left ventricular non-compaction (LVNC) is a rare form of congenital cardiomyopathy pathologically characterized by loose trabeculated myocardium and deep intertrabecular recesses or sinusoids, which is hypothesized to occur due to disruption of the compaction of loosely arranged myofibrils during development. The condition is estimated to have a population prevalence of less than 1 percent, but is notoriously difficult to accurately diagnose in life [11]. The association between LVNC and mitochondrial disorders stems from three main observations:

- the established association between LVNC and other genetically determined neuromuscular diseases (e.g., Barth syndrome, myotonic dystrophy 1 [*DMPK* mutations] and dysbrevinopathies [12];
- a small number of cases of Lebers hereditary optic neuropathy (LHON, m. 3460A>G), and MELAS (m.3243 A>G) in which a diagnosis of LVNC has been made;
- the detection of decreased mtDNA copy number and low-level rare (presumed somatic) heteroplasmic mtDNA point mutations in cardiac muscle from patients with idiopathic forms of LVNC [13].

Given the diagnostic challenges and rarity of both disorders, it remains unclear whether the lifetime risk of developing LVNC is significantly increased for patients with mitochondrial disorders. Future longitudinal studies in established cohorts of patients with mitochondrial disorders, together with improved cardiac imaging techniques (such as CMR), are likely to clarify whether LVNC is a true cardiac manifestation of mitochondrial disorders.

Electropathies

Bradyarrhythmias: The majority of bradyarrhythmias in mitochondrial disorders stems from disruption of the cardiac conduction system, with varying forms of atrio-ventricular block observed. Only a small number of patients have a form of bradyarrhythmia secondary to cardiovascular autonomic impairment [14].

There is a clear predilection for cardiac conduction defects in mtDNA deletion disorders such as KSS or CPEO, in comparison to those caused by point mutations or nuclear DNA mutations. Cardiac conduction defects are seen in up to 90 percent of patients with KSS encompassing PR-interval prolongation, second- or third-degree AV block, His-ventricular interval prolongation (due to distal disease) and Stokes Adams attacks [15], while tachyarrhythmias are extremely rare (see later in this chapter). This predilection may be due to the preferential accumulation of mtDNA deletions within the cardiac conduction system, though the molecular mechanisms mediating this remain unclear [16]. In contrast, less than 20 percent of patients with point mutations develop conduction defects, which, when present, commonly coexist with LV [17].

Tacchyarrythmias: Wolff-Parkinson-White syndrome (WPW) is a ventricular pre-excitation disorder, and the most common form of tachyarrhythmia observed in mitochondrial disorders. It has been observed in around 15 percent of patients with the m.8344A>G mutation [18] and in a similar proportion of those with the m.3243A>G in MELAS [1], compared to a population prevalence of roughly 1/1000. In patients with the m.3243A>G mutation, WPW usually develops before 40 years of age, and may precede the development of symptoms in other organ systems by more than a decade [19]. In our experience, this is one of the only instances where a cardiac manifestation of a mitochondrial disorder is likely to occur prior to the onset of symptoms in other organs. The risk of developing WPW does not appear to correlate with skeletal muscle heteroplasmy, though whether high levels of heteroplasmy within the myocardium mediate the development of this condition is unknown.

In addition, patients with Lebers hereditary optic neuropathy (LHON) may also have an increased risk of WPW, despite this being a condition not generally regarded as a multi-system phenotype [20]. These findings have yet to be replicated in large case series, but highlight that clinicians should remain vigilant for cardiac complications in all forms of mitochondrial disease [21].

Finally, a small number of cases of ventricular pre-excitation syndromes, such as ventricular tachycardia (VT), have been observed in patients with both mtDNA deletions and mtDNA point mutations.

Cardiac Involvement in Common Mitochondrial Disorders

Kearns-Sayre syndrome (KSS): Kearns-Sayre syndrome is characterized by: (1) symptom onset before 20 years, (2) a pigmentary retinopathy and (3) an ophthalmoparesis. Additional features such as cerebellar ataxia, deafness, dementia, a raised CSF protein and a cardiac conduction defect are also often present (see Chapter 3).

As mentioned, the cardiac conduction defects range from PR prolongation through to second- and third-degree block, his-ventricular block, DCM or Stoke-Adams attacks. The progression to third-degree heart block (complete heart block) can be rapid, and due to the combination of cognitive defects and reduced mobility that most patients exhibit, there may be no observable symptomatology in patients. This may explain how sudden death can occur in more than 10 percent of patients with KSS [22] and highlights the requirement for frequent cardiac monitoring in this cohort. Pacemaker implantation – while a vital intervention in patients with heart-block – may precipitate Torsade de points [23], and hence preventative implantable cardiac defibrillator (ICD) placement has been suggested by some centers, though it remains controversial.

m.3243A>G: The m.3243 mtDNA mutation was first described in MELAS, a multisystem heterogeneous mitochondrial syndrome characterized by (1) stroke-like episodes, (2) seizures, (3) dementia and (4) a myopathy with ragged red fibers (RRF) on muscle biopsy. As previously discussed, coexistent cardiac features are extremely common in m.3243A>G carriers, with approximately 50 percent of patients developing a cardiac abnormality. The most common cardiac complication is a hypertrophic cardiomyopathy without left ventricular outflow obstruction (LVOTO), which may later develop into a dilated cardiomyopathy. It appears that both the risk and severity of the cardiomyopathy are related to the myocardial mutant heteroplasmy load. WPW also develops in ~15 percent of patients, without clear association to underlying myocardial heteroplasmy level.

m.8344A>G: The m.8344A>m.8344A>G mutation was first described in myoclonus epilepsy and ragged red fibers (MERRF), and is characterized by myoclonus, epilepsy, ataxia and the finding of ragged red fibers on muscle biopsy. The cardiac manifestations are similar to those observed in m.3243A>G carriers, and are present in up to ~40 percent of patients. These include a cardiomyopathy and WPW in roughly equal proportions, with other cardiac abnormalities such as incomplete left bundle branch block present in only single cases [18]. The level of heteroplasmy required to cause a cardiomyopathy may be relatively low and possibly less than 15 percent. Consequently it is important for clinicians to remain vigilant in both known cases and their potentially affected relatives [24].

m.8993 T>G/C: The m.8993A>G/C mutation was first described in patients with neuropathy, ataxia and retinitis pigmentosa (NARP), which is a rare and phenotypically heterogeneous condition. Two mtDNA mutations (m.8993 T>G and m.8993 T>C) can cause either a NARP phenotype or Leigh-like syndrome, which is an often fatal infantile condition. Given the rarity of the genotype and the phenotypic heterogeneity, accurate quantification of the prevalence and phenotype of cardiac symptoms is difficult to determine, though cardiac failure remains a common cause of death in infants with Leigh syndrome [25]. This may be due to a tissue-specific predilection to alter mitochondrial function in the myocardium despite normal skeletal muscle function [26]. In contrast, some patients with NARP have no evidence of a cardiomyopathy or conduction defect approaching 60 years of age [27].

LHON: Leber's hereditary optic neuropathy (LHON) is a common mitochondrial disorder caused by homoplasmic point mutations that usually causes isolated dysfunction of the retinal ganglion cells (see Chapter 9). However, "LHON plus" syndromes (LHON with additional non-ophthalmological symptoms) exist, where cardiac manifestations such as myocardial hypertrophy, unexplained left ventricular dilatation [28] or ventricular pre-excitation disorders [20] may occur. Notably, however, these studies are from geographically isolated regions or single large extended pedigrees, raising the possibility that environmental or nuclear genetic modifiers may contribute to these findings [21].

Barth syndrome: Barth syndrome (alternatively called 3-methylglutaconic aciduria) is a rare multisystem mitochondrial disorder characterized by a dilated cardiomyopathy, skeletal myopathy, neutropenia and short stature. The disorder occurs due to loss-of-function mutations in the *TAZ* gene on the X-chromosome, and hence exclusively manifests in males with a frequency of approximately 1 in 400,000 live births [29].

The majority of individuals with Barth syndrome present at birth or in early infancy with hypotonia and a subsequent deceleration or delay in growth. Around 70 percent of boys also present with a cardiomyopathy within the first year of life, most commonly within the first six months [30]. The most common form of cardiomyopathy is a dilated cardiomyopathy, which is often accompanied by endocardial fibroelastosis (EFE) [31], and 50 percent have prominent left ventricular trabeculations in keeping with left ventricular non-compaction (LVNC). In contrast, hypertrophic cardiomyopathy has been described only in rare cases of Barth syndrome.

The acute presentation in early infancy, often in association with hypotonia, means that the differential diagnosis of Barth syndrome is broad, and may mimic viral or infectious etiologies, and hence the diagnosis is often delayed. However, as Barth syndrome accounts for ~5 percent of all forms of cardiomyopathy diagnosed in male infants, testing should be instigated in all appropriate subjects without delay by means of a diagnostic assay that measures the MLCL:L4-CL ratio on dried bloodspot cards [32].

Management of the cardiac complications of Barth syndrome involves standard medical therapy for heart failure, though there are no systematic studies of any particular treatment regime. Around 50 percent of boys will normalize their ejection fraction and left diastolic volumes with treatment [33], though pharmacological therapy is often necessitated throughout life even after the normalization of cardiac function. Long-term surveillance for cardiac arrhythmias should also be undertaken, with a low threshold for implantable cardiac defibrillators (ICDs) where symptomatology or electrophysiological investigations suggest a possible ventricular arrhythmia. Around one in seven children with Barth syndrome will require a cardiac transplantation, generally with good results [34], though a considered approach to immunosuppression must be undertaken given the coexistent neutropenia exhibited by patients.

Sengers syndrome: Sengers syndrome is a rare autosomal recessive multisystem mitochondrial disorder characterized by congenital cataracts, a hypertrophic cardiomyopathy, mitochondrial myopathy and an exertional lactic acidosis. The condition has recently been determined to result from homozygous or compound heterozygous mutations in the gene-encoding mitochondrial acylglycerol kinase (AGK) [35].

Fewer than 100 individuals have been described to date, but it appears that the condition can manifest most commonly as either a severe neonatal form caused by homozygous nonsense mutations or a more benign clinical course in which patients appear to have at least one mutation in a splice-site or start codon, and who may survive into their 40s [36].

In individuals with a severe form of disease, a severe hypertrophic cardiomyopathy is present at birth in almost call cases, with severe cardiac dysfunction from birth that invariably proves fatal in the first few months of life [36].

In contrast, the cardiac phenotype of individuals with the milder phenotype is not well described other than the consistent presence of a hypertrophic cardiomyopathy rather than a dilated cardiomyopathy in individuals. Additional cases are required in order to fully understand the cardiac phenotype of cases surviving into adulthood, and therefore we suggest patients are treated according to standard clinical management for hypertrophic cardiomyopathy.

Cardiac Investigations

All patients with any form of suspected or confirmed mitochondrial disease should have cardiac screening as part of their diagnostic evaluation and regularly throughout disease follow-up. Given the clinical heterogeneity of the cardiac complications (due to both the underlying genotype and inter-individual variation in heteroplasmy), determining the optimal frequency of cardiac surveillance and predicting the potential progression of cardiac disease are mired with difficulty. We therefore recommend that all physicians involved in patient care (e.g., neurologists, ophthalmologists) are familiar with basic surveillance investigations and, where indicated, a cardiologist familiar with mitochondrial disease is involved with patient follow-up and management.

Initial screening: Where a mitochondrial disease is confirmed or suspected, we recommend that affected individuals, as well as potential familial carriers, be screened with a standard 12-lead electrocardiogram (ECG) and 2-D echocardiography (ECHO) to enable a baseline functional and structural cardiac assessment (Figure 11.1). Where there is a phenotype with high risk of cardiac conduction defect (such as KSS), or in cases where WPW is possible (e.g., MELAS and MERRF), careful assessment of the ECG should be undertaken as emerging abnormalities may be subtle. Furthermore, given that ECHO is highly operator dependent, we would advise that it be undertaken in tertiary centers familiar with atypical cardiomyopathies.

Advanced imaging techniques: Cardiac MR and magnetic resonance spectroscopy (MRS) have largely been utilized in research settings in the mitochondrial patient population to date. Established findings include concentric LV hypertrophy, cardiac remodeling (increased torsion and longitudinal fiber shortening) [37] and impaired tissue bio-energetics (reduced PCr/ATP ratio) [38]. Abnormalities of myocardial function are also emerging in asymptomatic individuals with normal 12-lead ECG and ECHO who carry the m.3243A>G mutation [39].

Consequently, increasing interest is now also being paid to the utility of screening and surveillance with cardiac magnetic resonance imaging (CMR) and cardiac magnetic resonance spectroscopy. The techniques may offer an increased ability to detect structural abnormalities that may be missed with echocardiography (such as left ventricular non-compaction) and bio-energetic abnormalities of the myocardium that may act as diagnostic markers of mitochondrial disorders. As CMR findings correlate with urinary mutation load (itself a measure of disease severity), early CMR use – particularly in those most at risk of developing cardiac complications – would permit timely treatment commencement. However, at present, neither modality is routinely used for screening or surveillance.

More frequently at present, CMR is used to further define abnormalities identified using echocardiography. The improved sensitivity of CMR permits identification of characteristic cardiomyopathy patterns indicative of alternative genetic causes [40]. This is likely to be particularly important for cases where a molecular diagnosis has yet to be established, and such "deep phenotyping" of the myocardium may assist in establishing the molecular diagnosis in complex phenotypes in the setting of whole genome sequencing variant interpretation.

Cardiac biopsy: Cardiac biopsy is another potentially valuable investigation for select patients. However, as the risk of serious complication from cardiac biopsy is around 1 percent, this must be balanced against likely diagnostic yield. Therefore, as the molecular etiology in most patients with a phenotype suggestive of a mitochondrial disease will arise from (a) a pathogenic homoplasmic point mutation that is detectable in blood; (b) a nuclear DNA mutation that is detectable in blood; or (c) a heteroplasmic point mutation that is detectable in urine, skeletal muscle or blood, alternative, less-invasive investigative strategies are usually more appropriate.

In contrast, a cardiac biopsy may be useful in cases of isolated undiagnosed cardiomyopathies in which a mitochondrial diagnosis had not previously been considered. In these cases a combination of mitochondrial proliferation and evidence of impaired cytochrome c oxidase (COX) or succinate dehydrogenase (SDH) activity may be observed and prompt further investigation for a mitochondrial etiology (see Chapter 4). In our experience, however, this type of presentation is rare.

Cardiac Surveillance

Assuming that the initial screening investigations are normal in a patient, how often should they continue to undergo cardiac surveillance? While there are no specific guidelines, we advocate an approach similar to that suggested by Bates and colleagues [2] with annual follow-up for at least five years, extending to every three to five years where investigations (ECG and ECHO) are normal. When an abnormality is detected – either at baseline or during follow-up – the patient should be referred to a cardiologist with a subspecialty interest in mitochondrial disease for consideration of further investigation, pharmacological therapy or intervention as indicated (Figure 11.1).

Treatment

Cardiomyopathies: It has been suggested that treatment paradigms for mitochondrial cardiomyopathies follow those employed for HCM without LVOTO [2, 41]. These involve a combination of diuretics, B-blockers, angiotensin-converting (ACE) inhibitors and angiotensin receptor blockers (ARB) for patients with a low ejection fraction (< 50 percent). However, given the lack of clinical evidence for the treatment of cardiomyopathies in patients with mitochondrial disease specifically, their impact on slowing disease progression in this patient population is not clear. Given the lack of specific evidence, implantable cardioverter defibrillator (ICD) devices and cardiac resynchronization therapy should be used in line with standard guidelines [41].

Electrophysiological management of arrhythmias and conduction defects: There is some controversy over the management of arrhythmias and conduction defects arising in patients with mitochondrial disease. It has been suggested that permanent pacemaker (PPM) placement be undertaken in patients with mitochondrial disorders as soon as any evidence of atrioventricular block is observed (even first-degree heart block) [2] – especially in disorders such as KSS/CPEO in which progressive AV nodal disease is common and can be rapid. However, others suggest that conventional pacing/ICD implantation guidelines be followed [42]. We therefore advise the early involvement of a cardiologist with expertise in managing conduction defects in this patient population in the event of a mitochondrial patient developing relevant ECG changes.

Treatment for supra-ventricular tachyarrhythmias arising in those with mitochondrial diseases mirrors standard practice and incorporates both pharmacological and, where required, electrophysiological approaches. Similarly, the investigation of supra-ventricular tachycardias involves electrophysiological studies undertaken by subspecialist cardiologists to interrogate and ablate aberrant pathways where present.

Transplant: Cardiac transplantation is predominantly reserved for young patients with minimal noncardiac disease. However, in most patients, it is not easy to determine the likelihood of developing neurological or other significant systemic features, or the likely time course. Long-term follow-up data are lacking on the handful of patients with mitochondrial disease who have received cardiac transplants.

CASE REPORTS

Case Report 11.1

A 28-year-old male presented acutely following a collapse in a bar. He had been out with friends and had consumed approximately one bottle of wine. His friends described that he lost consciousness after climbing a flight of stairs and was unconscious for about three minutes. Earlier in the day, he had been complaining of shortness of breath and lethargy, which was new for him. He had no previous history of blackouts or shortness of breath.

Paramedics were called to the bar, and he was conscious on their arrival, with a blood pressure of 130/70, but a heart rate of 38/min. His ECG is shown in Figure 11.2.

On examination he remained bradycardic (heart rate 35), heart sounds were normal, the jugular venous pulse was not raised and there were no palpable heaves or thrills over his precordium. He was, however, noted to have mild non-fatigable bilateral ptosis, and stated that this had begun about a year ago, but he was not particularly concerned about it. Examination also revealed some diplopia on horizontal gaze bilaterally. Examination of his limbs showed a thin gentleman with low muscle mass and mild proximal weakness of the upper and lower limbs, but was otherwise normal.

Blood tests (FBC, U&Es, LFTs, CK, TFTs, calcium, phosphate and magnesium) were all normal (for abbreviations, see the list of abbreviations at the front of this volume).

Due to the conduction defect, the cardiologists performed an echocardiogram, which showed an ejection fraction of >50 percent and no evidence of systolic or diastolic dysfunction. They implanted a permanent pacemaker later that day.

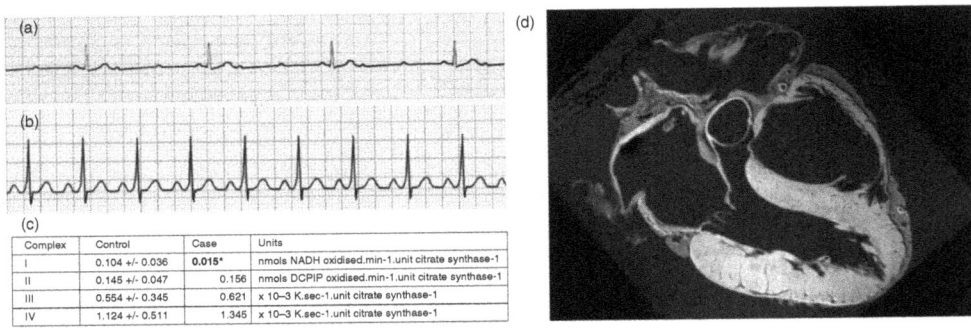

Complex	Control	Case	Units
I	0.104 +/- 0.036	0.015*	nmols NADH oxidised.min-1.unit citrate synthase-1
II	0.145 +/- 0.047	0.156	nmols DCPIP oxidised.min-1.unit citrate synthase-1
III	0.554 +/- 0.345	0.621	x 10–3 K.sec-1.unit citrate synthase-1
IV	1.124 +/- 0.511	1.345	x 10–3 K.sec-1.unit citrate synthase-1

Figure 11.2 ECGs and Mitochondrial functional studies for Case Reports 11.1–11.3
(a) Top – An ECG of the patient discussed in Case Report 11.1. (b) An ECG of the patient discussed in Case Report 11.2. (c) Mitochondrial enzymatic studies performed on the muscle biopsy of Case Report 11.3. (d) A cardiac MR image of the patient in Case Report 11.3.

Self-Assessment Questions

- What is the underlying rhythm abnormality?
- What do you suspect to be the most likely underlying mitochondrial condition?
- What molecular diagnostic investigations are likely to be helpful in order to determine the genetic etiology of his disease?
- Why are follow-up echocardiograms likely to be helpful?

Answers

- The ECG in this individual shows complete heart block. Note that the P waves show no relation to the QRS complexes. The atrial rhythm is regular at about 100 beats per minute (bpm), but the ventricular rate is approximately 40 bpm. A perfusing rhythm is maintained by a junctional or ventricular escape rhythm.
- The clinical features are suggestive (although not yet confirmed) of CPEO with ptosis and a probable extra-ocular gaze paresis bilaterally, though internal/central causes of this gaze palsy together with common mimics of CPEO (see Chapter 9) need to be thoroughly assessed on examination. In addition, the presence of a mild proximal myopathy is also consistent with a CPEO plus syndrome. Most single deletion disorders do not generally exhibit a family history (as in this case) occurring as *de novo* events resulting in ostensibly sporadic cases of disease in most families. Patients with CPEO are at high risk of developing conduction defects (as in this individual), and therefore the combination of his ocular abnormalities, probable myopathy and cardiac abnormalities at this stage point to CPEO, a mtDNA deletion disorder. In addition to a full clinical assessment (encompassing formal ophthalmological examination, neurological examination, respiratory function tests and EMG), molecular investigations for a potential mitochondrial disorder (probably CPEO+) will be required. Current best practice would recommend that a muscle biopsy with COX and SDH analysis (see Chapter 4) would be initially undertaken. Thereafter, long-range PCR and quantitative PCR should be undertaken to specifically look for a mtDNA deletion in this patient (see Chapter 4). In future, it is likely that whole genome sequencing techniques will detect deletions of mitochondrial DNA present in blood. However, it must be remembered that the lack of detection of mtDNA deletions from blood even in this instance could be falsely reassuring as: (a) the deletions may not be present in the individual's blood and be present only in other tissue types, in which case a muscle biopsy will still be required, and: (2) the threshold for the detection of mtDNA deletions using next-generation sequencing techniques may limit their utility to exclude mtDNA deletions when present in blood but below the level of detection. As with all clinical tests, therefore, clinicians using them should be aware of their specific limitations.
- Echocardiograms are going to be of significant utility in this patient given the high likelihood that he will also develop a cardiomyopathy (~60 percent). Careful surveillance will enable timely intervention with appropriate pharmacotherapy should this arise.

Approach and Clinical Management

A full clinical examination did indeed reveal that his ocular features were in keeping with CPEO, and phenomimics were excluded (see Chapters 7 and 9).

A muscle biopsy was subsequently performed showing a mosaic deficiency of COX and SDH, and long-range PCR and quantitative PCR studies revealed a ~ 5kb deletion of mtDNA in his muscle that were present in relatively high levels (~50 percent). He was referred to ophthalmology for correction of his ptosis, and the cardiology team, satisfied with his pacemaker function, arranged yearly echocardiograms and pacemaker checks. Annual blood sugar and HbA_1C monitoring were arranged through the neurology department in light of the risk of developing diabetes.

Case Report 11.2

A 22-year-old female was seen in the cardiology outpatient clinic with palpitations. She had experienced several recent similar episodes that were all self-limiting and lasted up to one hour. She had taken her pulse during the attacks and thought it had been more than 200 beats per minute. She had been symptomatic with shortness of breath but had not lost consciousness.

Her other medical problems included myoclonic epilepsy, which was diagnosed at age 10, though she had not attended follow-up since her early teens. Her seizures had recently changed from a myoclonic pattern to include generalized tonic-clonic seizures, and she was therefore waiting to see the neurology team again. She also had some long-standing severe hearing impairment, but she was not sure of the exact cause.

Bedside cardiovascular examination was entirely normal. Her neurological examination revealed a degree of cognitive impairment. She exhibited a mild and predominantly mid-line ataxia, mild proximal weakness and some intermittent myoclonus. Ophthalmological examination was normal.

She took sodium valproate (800 mg BD) and denied any drug or alcohol excess. Blood tests (FBC, U&Es, LFTs, magnesium, calcium, TFTs and serum sodium valproate levels) were all normal. Her resting ECG is shown in Figure 11.2.

Her estranged mother had a "mitochondrial disease," though she was not sure which one, but she was certain that she had not had any heart problems.

Self-Assessment Questions

- What does the ECG show?
- What is the most likely underlying mitochondrial syndrome?
- What are the most likely mtDNA mutations causing this syndrome and how is it easiest to test for this?
- What is the appropriate cardiological management for this individual?

Answers

- The ECG shows characteristic features of Wolff-Parkinson White (WPW) syndrome. The PR interval is less than 120 ms, a delta wave (slurring rise of the initial portion of the QRS complex) is present and QRS prolongation beyond 110 ms can be seen. Although this patient does not exhibit it, a "pseudo-infarction" pattern can also be seen in ~ 70 percent of patients and in which a deflected delta wave in the inferior and anterior leads, or prominent R waves in V1–V3 can be seen.
- While generally considered a static ECG finding, some individuals show ECG features only intermittently, and in some patients, the features of pre-excitation may become

more pronounced with maneuvers that raise vagal tone. In addition, patients may possess a "retrograde-only" pathway of WPW. Here, all anterograde conduction goes though the AV node, so none of the classical ECG features of WPW is seen. The frequency of this form of WPW in mitochondrial disorders is unknown, but highlights the importance of a thorough clinical history alongside surveillance investigations such as an ECG. A low index of suspicion of any form of rhythm abnormality should prompt discussion with a cardiologist, even when an ostensibly normal ECG is present.

- The most likely diagnosis here is MERRF. The finding of myoclonic epilepsy (that from the clinical history appears to be beginning to generalize) is very unusual in other forms of mitochondrial disease (see Chapters 2, 3 and 7), and this would be the most likely diagnosis in this individual.

- The m.8344A>G mutation is responsible for at least 80 percent of all cases of MERRF. A further three variants (m.8356 T>C, m.8363 G>A and m.8361 G>A) are responsible for a further 10 percent of cases. An additional 10 percent of mutations are found in other genes.

- Previous diagnostic guidelines suggest to simply test for the m.8344A>G mutation initially in blood or muscle. However, with reducing costs of high-depth sequencing, whole mitochondrial genome sequencing in blood is likely to be the most appropriate first-line test in coming years. This approach may assist in detecting the additional 10 percent of genetic variants.

- This patient requires an echocardiogram in the first instance to look for any signs of left ventricular impairment. She will also require a referral to a cardiologist with an expertise in electrophysiology even if the echocardiogram is normal. She is likely to require electrophysiological studies (EPS) and may benefit from radiofrequency ablation.

Approach and Management

This individual had whole mtDNA genome sequencing looking for point mutations performed on a blood sample, detecting the m.8344A>G mutation with 24 percent heteroplasmy. The same variant was sequenced in DNA extracted from the muscle biopsy revealing 56 percent heteroplasmy. The diagnosis of MERRF was thus confirmed and she was referred to the cardiologists who performed EPS and radiofrequency ablation. Her echocardiogram was normal. She remained asymptomatic over the next year, and was discharged from cardiology follow-up with a plan for ongoing surveillance to be conducted by the neurology team looking after her. This involves ongoing yearly echocardiograms and ECGs.

Case Report 11.3

A 32-year-old male with a lifelong neurodevelopmental disability (diagnosed as a "leukoencephalopathy of unknown cause") and who is resident in a care home was admitted to the Medical Admissions Unit with shortness of breath.

He had an additional medical history that included a diagnosis of asthma and diabetes controlled by insulin. The care home staff described that his shortness of breath had been getting progressively worse for the past three weeks. He had taken a course of steroids and antibiotics from his general practitioner with little effect for a presumed infective exacerbation of his asthma. On further questioning, the care home staff said that he had also been progressively lethargic, and he himself agreed that he was short of breath lying flat, though he could not offer much more history. His exercise tolerance was now only approximately 20 yards, in contrast to being able to walk a mile around a year ago.

On examination he was noted to be of short stature (5'3", 160 cm), had a raised JVP, together with bi-basal crackles on auscultation of his chest and some pitting pedal edema. There were no murmurs on auscultation. Neurologically he had profound cognitive impairment and widespread upper motor neuron signs together with a significant degree of muscle wasting.

Blood tests (FBC, U&Es, LFTs, CRP, CK, magnesium, calcium and phosphate) were normal. An ECG was normal, but a transthoracic echo showed severe global ventricular hypertrophy, with an ejection fraction of 25 percent. There was no evidence of any outflow obstruction or valvular abnormalities. A diagnosis of a hypertrophic cardiomyopathy was made.

He went on to have an endomyocardial biopsy to look for histological evidence of HCM. Instead, this revealed marked interstitial fibrosis and hypertrophic cardiomyocytes, together with a proliferation hypertrophy of cardiomyocytes and a mosaic deficiency of COX and SDH, which for the first time alerted the clinicians to the possibility of a mitochondrial disorder. Further immunohistochemical analysis showed a profound deficit in complex I of the mitochondrial respiratory chain (Figure 11.2).

Self-Assessment Questions

- What further investigations (clinical and molecular) should be performed?
- Do you think this individual should have had a cardiac biopsy?
- What are the possible treatment options available for this patient?

Answers

- This patient has an undiagnosed mitochondrial disorder. While the phenotype may appear clearly suggestive of a mitochondrial disease to some, it is quite common for patients to have a constellation of symptoms indicative of a mitochondrial disorder but go many years without a diagnosis. This often occurs as patients are managed by specialties where symptoms are observed and managed in isolation. This patient therefore requires a clinical ophthalmological examination, neurological examination, repeat neurological imaging (preferably MRI of his brain), glucose and HbA1c as a minimum.
- Whilst the differential diagnosis remains broad even within the sphere of mitochondrial disorders, conditions such as m.3243A>G/MELAS remain high on the list of differential diagnoses given the leukoencephalopathy, deafness, diabetes and short stature.
- A muscle biopsy is not now required as the salient biochemical information can be taken from the cardiac muscle biopsy (providing their is frozen, not fixed tissue available for molecular analysis). In the first instance, screening for pathogenic mtDNA mutations in blood and from DNA extracted from the myocardium should be performed.
- The cardiac biopsy was undertaken by the clinical team to identify the etiology of his cardiomyopathy, and until this point they had not considered a mitochondrial disorder. It was therefore entirely reasonable, and led to the diagnosis. Had the possibility of a

Table 11.1 A table of the cardiac phenotypes of patients with the most common mtDNA disorders with a cardiological abnormality as part of their phenotype

Syndrome	Example genotypes	Cardiomyopathy			Electropathy	
		Hypertrophic cardiomyopathy	Left ventricular non-compaction	Dilated	Ventricular pre-excitation	Conduction defect
MELAS	m.3243A>G	++	+	+	++	+
MERRF	m.8344A>G	++	-	-	++	+
CPEO/KSS	Single large deletions (~5kb)	+	+	+	-	++

The key clinical syndromes and the most common genotypes causing these diseases are shown. ++ represents a phenotype observed in >1 publication, with + representing single case reports plus the author's experience. Key – MELAS (mitochondrial encephalopathy with lactic acidosis and stroke-like episodes), MERRF (myoclonic epilepsy with ragged red fibers), CPEO (chronic progressive external ophthalmoplegia), KSS (Kearns-Sayre syndrome).

mitochondrial disorder been raised previously, then a skeletal muscle biopsy would have been the investigation of choice to look for biochemical evidence of mitochondrial dysfunction, and would most likely have been abnormal. It is important to note that any tissue taken at the time of cardiac biopsy should be frozen and fixed to prevent the necessity of a repeat biopsy in a patient.

- This patient has a severe cardiomyopathy. He should be referred as an emergency for both stabilization of his acute LV impairment and long-term management of his cardiomyopathy. Given he is symptomatic and has severe LV impairment, this may well involve not only pharmacological therapy but also cardiac resynchronization therapy.

Approach

In this patient, an MRI scan of his brain was suggestive of m.3243A>G, which was confirmed in both his blood (7 percent heteroplasmy) and in his myocardium (77 percent heteroplasmy) (Figure 11.2). His immediate treatment under the cardiology team was to stabilize his left ventricular failure and instigate pharmacological therapy. In parallel, his family were contacted in order to attend the neurology clinic for assessment and cardiac screening.

References

1. Anan R, Nakagawa M, Miyata M, et al. Cardiac involvement in mitochondrial diseases. A study on 17 patients with documented mitochondrial DNA defects. *Circulation* 1995; 91(4): 955–961.

2. Bates MG, Bourke JP, Giordano C, d'Amati G, Turnbull DM, Taylor RW. Cardiac involvement in mitochondrial DNA disease: Clinical spectrum, diagnosis, and management. *European Heart Journal* 2012; 33(24): 3023–3033.

3. Limongelli G, Tome-Esteban M, Dejthevaporn C, Rahman S, Hanna MG, Elliott PM. Prevalence and natural history of heart disease in adults with primary mitochondrial respiratory chain disease. *European Journal of Heart Failure* 2010; 12 (2): 114–121.

4. Florian A, Ludwig A, Stubbe-Drager B, et al. Characteristic cardiac phenotypes are detected by cardiovascular magnetic resonance in patients with different clinical phenotypes and genotypes of mitochondrial myopathy. *Journal of Cardiovascular Magnetic Resonance: Official Journal of the Society for Cardiovascular Magnetic Resonance* 2015; 17: 40.

5. Taylor RW, Giordano C, Davidson MM, et al. A homoplasmic mitochondrial transfer ribonucleic acid mutation as a cause of maternally inherited hypertrophic cardiomyopathy. *Journal of the American College of Cardiology* 2003; 41(10): 1786–1796.

6. Hagen CM, Aidt FH, Havndrup O, et al. Private mitochondrial DNA variants in Danish patients with hypertrophic cardiomyopathy. *PloS One* 2015; 10(4): e0124540.

7. Vydt TC, de Coo RF, Soliman OI, et al. Cardiac involvement in adults with m.3243A>G MELAS gene mutation. *The American Journal of Cardiology* 2007; 99 (2): 264–269.

8. Okajima Y, Tanabe Y, Takayanagi M, Aotsuka H. A follow up study of myocardial involvement in patients with mitochondrial encephalomyopathy, lactic acidosis, and stroke-like episodes (MELAS). *Heart* 1998; 80(3): 292–295.

9. Tveskov C, Angelo-Nielsen K. Kearns-Sayre syndrome and dilated cardiomyopathy. *Neurology* 1990; 40(3 Pt. 1): 553–4.

10. Stalder N, Yarol N, Tozzi P, et al. Mitochondrial A3243 G mutation with

manifestation of acute dilated cardiomyopathy. *Circulation Heart Failure* 2012; 5(1): e1–3.

11. Kohli SK, Pantazis AA, Shah JS, et al. Diagnosis of left-ventricular non-compaction in patients with left-ventricular systolic dysfunction: Time for a reappraisal of diagnostic criteria? *European Heart Journal* 2008; 29(1): 89–95.

12. Finsterer J. Cardiogenetics, neurogenetics, and pathogenetics of left ventricular hypertrabeculation/noncompaction. *Pediatric Cardiology* 2009; 30(5): 659–681.

13. Tang S, Batra A, Zhang Y, Ebenroth ES, Huang T. Left ventricular noncompaction is associated with mutations in the mitochondrial genome. *Mitochondrion* 2010; 10(4): 350–357.

14. Di Leo R, Musumeci O, de Gregorio C, et al. Evidence of cardiovascular autonomic impairment in mitochondrial disorders. *Journal of Neurology* 2007; 254(11): 1498–1503.

15. Young TJ, Shah AK, Lee MH, Hayes DL. Kearns-Sayre syndrome: A case report and review of cardiovascular complications. *Pacing and Clinical Electrophysiology: PACE* 2005; 28(5): 454–457.

16. Muller-Hocker J, Jacob U, Seibel P. The common 4977 base pair deletion of mitochondrial DNA preferentially accumulates in the cardiac conduction system of patients with Kearns-Sayre syndrome. *Modern Pathology: An Official Journal of the United States and Canadian Academy of Pathology, Inc.* 1998; 11(3): 295–301.

17. Majamaa-Voltti K, Peuhkurinen K, Kortelainen ML, Hassinen IE, Majamaa K. Cardiac abnormalities in patients with mitochondrial DNA mutation 3243A>G. *BMC Cardiovascular Disorders* 2002; 2: 12.

18. Wahbi K, Larue S, Jardel C, et al. Cardiac involvement is frequent in patients with the m.8344A>G mutation of mitochondrial DNA. *Neurology* 2010; 74(8): 674–677.

19. Sproule DM, Kaufmann P, Engelstad K, Starc TJ, Hordof AJ, De Vivo DC. Wolff-Parkinson-White syndrome in patients with MELAS. *Archives of Neurology* 2007; 64(11): 1625–1627.

20. Nikoskelainen EK, Savontaus ML, Huoponen K, Antila K, Hartiala J. Pre-excitation syndrome in Leber's hereditary optic neuropathy. *Lancet* 1994; 344(8926): 857–858.

21. Kirkman MA, Yu-Wai-Man P, Korsten A, et al. Gene-environment interactions in Leber hereditary optic neuropathy. *Brain: A Journal of Neurology* 2009; 132(Pt 9): 2317–2326.

22. Khambatta S, Nguyen DL, Beckman TJ, Wittich CM. Kearns-Sayre syndrome: A case series of 35 adults and children. *International Journal of General Medicine* 2014; 7: 325–332.

23. Subbiah RN, Kuchar D, Baron D. Torsades de pointes in a patient with Kearns-Sayre syndrome: A fortunate finding. *Pacing and Clinical Electrophysiology: PACE* 2007; 30 (1): 137–139.

24. DiMauro S, Hirano M. Merrf. In Pagon RA, Adam MP, Ardinger HH, et al., eds. GeneReviews(R). Seattle (WA); 1993.

25. Thorburn DR, Rahman S. Mitochondrial DNA-associated Leigh Syndrome and NARP. In Pagon RA, Adam MP, Ardinger HH, et al., eds. GeneReviews(R). Seattle (WA); 1993.

26. Thorburn DR, Chow CW, Kirby DM. Respiratory chain enzyme analysis in muscle and liver. *Mitochondrion* 2004; 4(5–6): 363–375.

27. Rawle MJ, Larner AJ. NARP Syndrome: A 20-Year Follow-Up. *Case Reports in Neurology* 2013; 5(3): 204–207.

28. Sorajja P, Sweeney MG, Chalmers R, et al. Cardiac abnormalities in patients with Leber's hereditary optic neuropathy. *Heart* 2003; 89(7): 791–792.

29. Dudek J, Maack C. Barth syndrome cardiomyopathy. *Cardiovascular Research* 2017.

30. Roberts AE, Nixon C, Steward CG, et al. The Barth Syndrome Registry: Distinguishing disease characteristics and growth data from a longitudinal study.

American Journal of Medical Genetics Part A 2012; 158A(11): 2726–2732.

31. Clarke SL, Bowron A, Gonzalez IL, et al. Barth syndrome. *Orphanet J Rare Dis* 2013; 8: 23.

32. Kulik W, Van Lenthe H, Stet FS, et al. Bloodspot assay using HPLC-tandem mass spectrometry for detection of Barth syndrome. *Clin Chem* 2008; 54(2): 371–378.

33. Spencer CT, Bryant RM, Day J, et al. Cardiac and clinical phenotype in Barth syndrome. *Pediatrics* 2006; 118(2): e337–46.

34. Mangat J, Lunnon-Wood T, Rees P, Elliott M, Burch M. Successful cardiac transplantation in Barth syndrome – single-centre experience of four patients. *Pediatr Transplant* 2007; 11(3): 327–331.

35. Mayr JA, Haack TB, Graf E, et al. Lack of the mitochondrial protein acylglycerol kinase causes Sengers syndrome. *American Journal of Human Genetics* 2012; 90(2): 314–320.

36. Haghighi A, Haack TB, Atiq M, et al. Sengers syndrome: Six novel AGK mutations in seven new families and review of the phenotypic and mutational spectrum of 29 patients. *Orphanet J Rare Dis* 2014; 9: 119.

37. Hollingsworth KG, Gorman GS, Trenell MI, et al. Cardiomyopathy is common in patients with the mitochondrial DNA m.3243A>G mutation and correlates with mutation load. *Neuromuscular Disorders: NMD* 2012; 22(7): 592–596.

38. Lodi R, Rajagopalan B, Blamire AM, Crilley JG, Styles P, Chinnery PF. Abnormal cardiac energetics in patients carrying the A3243 G mtDNA mutation measured in vivo using phosphorus MR spectroscopy. *Biochim Biophys Acta* 2004; 1657(2–3): 146–150.

39. Bates MG, Hollingsworth KG, Newman JH, et al. Concentric hypertrophic remodelling and subendocardial dysfunction in mitochondrial DNA point mutation carriers. *Eur Heart J Cardiovasc Imaging* 2013; 14(7): 650–658.

40. Partington SL, Givertz MM, Gupta S, Kwong RY. Cardiac magnetic resonance aids in the diagnosis of mitochondrial cardiomyopathy. *Circulation* 2011; 123(6): e227–9.

41. Authors/Task Force m, Elliott PM, Anastasakis A, et al. 2014 ESC guidelines on diagnosis and management of hypertrophic cardiomyopathy: The Task Force for the Diagnosis and Management of Hypertrophic Cardiomyopathy of the European Society of Cardiology (ESC). *European Heart Journal* 2014; 35(39): 2733–2779.

42. Brignole M, Auricchio A, Baron-Esquivias G, et al. 2013 ESC guidelines on cardiac pacing and cardiac resynchronization therapy: The Task Force on Cardiac Pacing and Resynchronization Therapy of the European Society of Cardiology (ESC). Developed in collaboration with the European Heart Rhythm Association (EHRA). *European Heart Journal* 2013; 34 (29): 2281–2329.

Respiratory Medicine and Anesthesiology

Philip G. Morgan and Margaret M. Sedensky

Introduction

Children with mitochondrial disease often present with a constellation of signs and symptoms indicative of systemic disease [1]. Respiratory insufficiency can accompany mitochondrial disease, and is a serious consequence of mitochondrial dysfunction. Seldom a sole finding, respiratory insufficiency is attributable both to the myopathy that is common among these patients and to the encephalopathy that is a hallmark of mitochondrial disease. In addition, prolonged myopathies can profoundly change the spine or chest wall, leading to restrictive lung disease. Any medications or conditions that contribute to muscle weakness, decreased chest excursion or central respiratory depression can potentially lead to respiratory failure in mitochondrial patients [2]. These patients often become hypercarbic during sleep, and require continuous positive airways pressure (CPAP) or bilevel positive airway pressure (BIPap) overnight. When patients with mitochondrial disease undergo an acute insult, such as a viral illness, the inadequacy of the muscles of respiration to sustain adequate ventilation is often the beginning of a downward spiral in their clinical course. Finally, decreased gastric motility and defective swallowing may both lead to an increased risk of aspiration, a potentially lethal respiratory complication in these patients.

Respiratory Disease – Clinical Presentation and Management

Decreased respiratory strength: Mitochondrial disease is a well-known cause of myopathy with resulting weakness of all muscles including the respiratory muscles of the thorax. Respiratory failure resulting from mitochondrial disease can present in the neonatal period or later in life [3, 4]. In addition, these defects can affect the diaphragm, which has even been noted in neonates with mitochondrial dysfunction [5]. In rare cases, this weakness can progress very rapidly, resulting in fulminant respiratory failure [6]. Supportive care, including positive pressure ventilation, is often necessary.

Restrictive lung disease: Mitochondrial defects do not directly cause restrictive lung disease. However, prolonged myopathies, including mitochondrial myopathies, are well known to lead to scoliosis and kyphosis, both of which are causes of restricted chest excursion. In severe cases, patients can present with arthrogryposis with symptoms affecting the chest as well as other joints [3, 7]. As a result, it is not unusual for these patients to present for scoliosis repair with pronounced restrictive disease. This is, of course, compounded by the myopathic weakness and potential decrease in central respiratory drive. As a result, these patients often need intensive pulmonary support postoperatively and in the presence of pulmonary infections.

Pulmonary hypertension: Mitochondrial disease is a rare cause of pulmonary hypertension [8]. However, pulmonary hypertension has been reported to be caused by point mutations in mitochondrial DNA and by mutations in the iron sulfur protein *NFU1* [9]. It was suggested that mitochondrial disease should be considered when patients have coexisting encephalopathy and pulmonary hypertension. The prognosis in such cases is poor.

Central respiratory depression: Decreased respiratory drive can be a result of the encephalopathy that is a sign of mitochondrial disease. It is often a hallmark of a particular type of mitochondrial dysfunction that leads to an acidotic state called Leigh syndrome. Leigh syndrome patients whose disease is caused by mutation in a gene called *SURF1* may be most susceptible to decreased central respiratory drive [10]. Mitochondrial disease leading to central respiratory depression is also seen in animal models of complex I dysfunction [11]. Many cases of Leigh syndrome show defects in complex I, an entry point into the electron transport chain.

Aspiration: Aspiration is a serious potential risk in mitochondrial patients as in any patient with myopathy or swallowing abnormalities [12]. However, mitochondrial patients face the additional problem of potentially being unable to meet the metabolic demands necessary to fight the infection, potential fever and tachypnea or increased work of breathing that is a consequence of aspiration. Aspiration is a potentially fatal complication that must be avoided in these patients [13].

As with other manifestations of mitochondrial disease, there is only supportive care of the respiratory system in these patients. Excellent pulmonary toilet and meticulous care of concurrent infections are both key. Since infections can lead to hyperthermia, tachypnea, tachycardia, increased work of breathing and overall increase in metabolic demand, it is key to absolutely minimize the risk of infection during any hospitalization. Likewise, during a hospitalization, any postsurgical stress from pain or wound healing must be minimized. Overall a careful fluid balance must take into account any temperature elevation, generally poorly tolerated in these children, as well as tachypnea due to pain or fever. It is important to remember that these patients may have little pulmonary reserve with which to respond to increased metabolic demand.

Anesthetic Considerations for Children with Mitochondrial Disease

Preoperative care It is not uncommon for children with mitochondrial disease to present for surgical treatment of concomitant conditions [14]. In general, these periods of increased stress will increase metabolic burden on patients with poor mitochondrial function. Young children require a general anesthetic for almost any potentially painful procedure, and virtually all anesthetic medications are gastrointestinal, cardiac and respiratory depressants [2].

Many of these patients have experienced multiple operative procedures. Parents of chronically ill patients are often very attuned to their child's needs during such hospitalizations, and know the details of how they did with past surgical procedures and anesthetics. Many are concerned about prolonged hospitalization or ventilatory support, loss of developmental milestones or prolonged nausea and vomiting after an anesthetic. Surgical procedures can be very short, like a muscle biopsy for diagnosis of disease, or very lengthy and accompanied by significant fluid shifts, like repairing a scoliotic spine.

As with any preoperative evaluation, a good history and very recent physical examination are mandatory with emphasis on cardiac, pulmonary, gastrointestinal and central nervous systems [4, 15]. These patients often change dramatically in a relatively short time, so such exams should be performed close to the time of surgery. However, time should be available for appropriate laboratory or diagnostic tests. Generally, within one to two weeks prior to the surgery is adequate to ensure good care (Figure 12.1). In general, the following systems need special preoperative measures:

- *Fluids*: Patients with mitochondrial disease do not tolerate prolonged fasting very well [2, 14]. Elective cases are best served as first starts of the day, with preoperative hydration and minimization of preoperative fasting. It is often adequate to hydrate these patients orally. If the procedure cannot be done early in the day, the patient's period of no oral feedings should be adjusted to be two to three hours prior to surgical start time if the particular procedure allows. Restriction of oral intake must be tailored to the constraints of the particular surgical requirements. For example, surgeries of the gastrointestinal system may require prolonged fasting such that a preoperative admission for insertion of an intravenous line and intravenous hydration is appropriate. In general, when long periods of fasting are required (e.g., gastrointestinal procedures), discussion with the proceduralist and the anesthesiologist is mandatory to maximize care for the patient and determine whether preoperative admission for intravenous fluids is indicated.

  The patients should not be rendered hypoglycemic, and many will be optimally maintained on an intravenous glucose-containing solution. In contrast, many children with refractory seizures will be maintained on ketogenic diets and should not have glucose- or lactate-containing solutions perioperatively. A reasonable strategy is to provide glucose at a maintenance rate to patients on a glucose-containing diet, with regular monitoring of glucose levels. Patients on ketogenic diets require similar amounts of fluid, but without glucose in the solution. Exact fluid management should be discussed with a primary physician who chronically cares for these patients.

  It is often stated that no mitochondrial patients should be given lactated Ringer's (LR) solution, which contains 28 mmol of lactate per liter of fluid. Theoretically, this makes sense since many of these patients will have difficulty metabolizing the lactate. However, it is unlikely that a short exposure of these patients to a small amount of lactate will present a serious problem. The gravest concern is that a patient may receive a large amount of LR even in a short procedure, or that the patient may receive the LR for a prolonged period. To avoid these possibilities, it is probably best to avoid LR entirely. However, for short procedures, or in a critical situation with no other fluid available, most patients can tolerate small amounts of lactate.

- *Cardiac*: The status of the heart must be well understood preoperatively and its function optimized, since general anesthetics can have profound effects on the myocardium. Conduction defects and cardiomyopathies would limit use of medicines that can cause arrhythmias or worsen cardiac function. For example, patients with poor cardiac function may be anesthetized with a technique that relies primarily on opioids that do not depress the myocardium (as opposed to volatile anesthetics, see later in this chapter). Most institutions will require cardiac evaluation of mitochondrial patients annually or semiannually, especially if a surgical procedure is anticipated. It is best to plan this relatively early in the preoperative period so that any adjustments to cardiac care may be made with adequate advance notice.

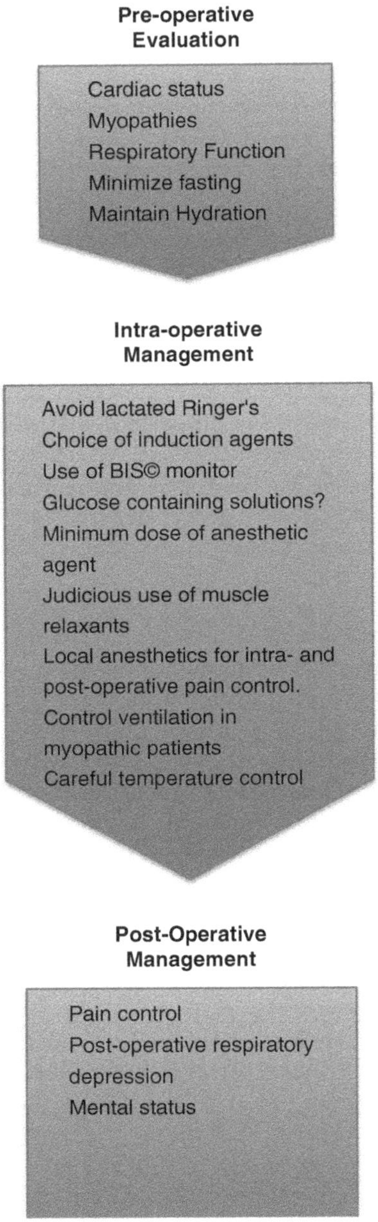

Pre-operative Evaluation

Cardiac status
Myopathies
Respiratory Function
Minimize fasting
Maintain Hydration

Figure 12.1 Major considerations during the perioperative period in patients with mitochondrial dysfunction. The decisions concerning each of these points may vary with individual patients. However, they each should be considered with all such patients and are discussed in the text.

Intra-operative Management

Avoid lactated Ringer's
Choice of induction agents
Use of BIS© monitor
Glucose containing solutions?
Minimum dose of anesthetic agent
Judicious use of muscle relaxants
Local anesthetics for intra- and post-operative pain control.
Control ventilation in myopathic patients
Careful temperature control

Post-Operative Management

Pain control
Post-operative respiratory depression
Mental status

- *Respiratory*: Respiratory depression and blunting of the normal response to either hypoxia or hypercarbia are well-known side effects of most anesthetics and some analgesics, like opioids. However, even anesthetics and sedatives with minimal respiratory effects under normal conditions may lead to profound loss of respiratory drive in mitochondrial patients. As a general rule, it is best to minimize preoperative sedative exposure to only that absolutely necessary. Respiratory depression in mitochondrial patients is the result of a combination of inhibited central respiratory drive, decreased muscle strength for the work of breathing and potentially airway obstruction due to decreased motor tone of pharyngeal muscles. It is often easier to address the latter two issues than the decrease in central respiratory drive. If present, ongoing dysphagia must be noted and care paid to aspiration risks. This is especially true as decreased gastric motility has been reported in patients with mitochondrial dysfunction.

Intraoperative Care

- *Anesthetics*: Choice of general anesthetic regimes can be a challenge, even for the most experienced anesthesiologist. There are reports of unexpected, devastating outcomes from cases in which the anesthetic management was routine and uneventful during the surgical procedure [16]. Most significant are reports of delayed white matter degeneration or profound respiratory depression. However, almost any type of anesthetic protocol has been used safely and effectively for mitochondrial patients. In general, cardiac and respiratory depression from general anesthesia is the uppermost concern for these medications, as is muscle relaxation from paralytic agents. Since mitochondrial diseases encompass a spectrum of multisystem disorders, customization of anesthetic management to the patient's condition is, of course, the rational way to proceed.

- *Volatile anesthetics*: It has been demonstrated that volatile anesthetics as a group directly depress mitochondrial function, especially the activity of complex I and complex V (Figure 12.2) [17, 18]. In addition there are interesting data to suggest that a certain type of mitochondrial defect renders patients very sensitive to volatile anesthetics. For patients with a known diagnosis of complex I dysfunction, it may be prudent to carefully monitor depth of anesthesia. In a small cohort of children it was observed that children with this particular form of disease were very hypersensitive to a volatile agent called sevoflurane [19]. In our practice we try to use the lowest dose of a volatile anesthetic necessary for unconsciousness and analgesia. We commonly see that patients who previously were very slow to recover from a volatile anesthetic have much shorter recovery times with this precaution in mind. We therefore routinely use a BIS© monitor, a commercial device that displays in real time a number that is used as a surrogate for depth of anesthesia, in all patients for whom we suspect hypersensitivity to anesthetics.

  All the volatile anesthetics cause respiratory and cardiac depression, as well as muscle weakness. However, each has a unique profile in these other effects. For example isoflurane and desflurane maintain cardiac output better than sevoflurane, but depress ventilatory drive to CO_2 more than sevoflurane. None of the more recently developed volatile anesthetics is metabolized to a significant degree, and therefore can be considered to be excreted unchanged through the lungs. This is a distinct advantage over the many intravenous agents that must be metabolized by the liver and excreted through the kidney to be cleared from the patient.

- *Intravenous anesthetics*: Intravenous medications as part of an anesthetic protocol include drugs like propofol, etomidate, ketamine and pentobarbital and opioids such as morphine and fentanyl. Many of these intravenous medications, including propofol, have been shown to directly inhibit mitochondria primarily at complex I (Figure 12.2). Propofol is probably the most promiscuous of the IV agents in its ability to depress multiple steps of mitochondrial metabolism [20]. There has been long-standing controversy surrounding propofol as a suitable agent for mitochondrial patients, since

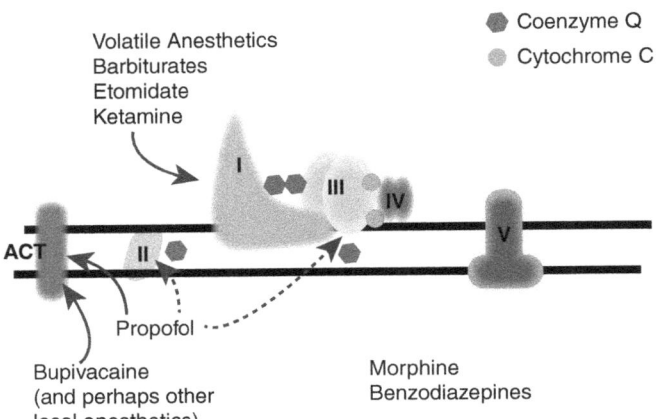

Figure 12.2 The primary sites at which common anesthetic agents affect mitochondrial function. The two most notable agents are a) the volatile anesthetics, which are strong inhibitors of complex I function; and b) propofol, which is reported to affect multiple sites. However, the primary site of propofol action in mitochondria is felt to be inhibition of the transport of fatty acids into the matrix. Morphine and benzodiazepines are without arrows since their effects are inconsistent in different studies. (A black and white version of this figure will appear in some formats. For the color version, please refer to the plate section.)

a syndrome characterized by profound lactic acidosis and elevation of acylcarnitines has been associated with prolonged infusion of this medication (propofol infusion syndrome). However, propofol, like volatile anesthetics, has also been used safely in many patients with mitochondrial disease [2, 14].

- *Muscle relaxants*: Muscle relaxants do not inhibit mitochondrial function, but certainly contribute to residual muscle weakness, and therefore are of concern for postoperative respiratory insufficiency, dysphagia and weakness. Children with mitochondrial disease who present with hypotonia usually require little, if any, muscle relaxation for most surgical procedures.

  Patients with preexisting myopathies are very sensitive to muscle relaxants such as atracurium, rocuronium or pancuronium. These are all termed nondepolarizing muscle relaxants and do not warrant concern other than their potentially profound effects on a hypotonic patient. However, another drug, succinylcholine, depolarizes muscle and has a special complication of increasing potassium levels in patients with certain kinds of myopathies. Therefore it is generally avoided, if possible, in patients with mitochondrial disease, although there is a paucity of data to support this concern. It is also important to note that mitochondrial disease was previously thought to place patients at increased risk for malignant hyperthermia, a life-threatening hypermetabolic response to volatile anesthetics and succinylcholine. However, further study has failed to support this association. Patients with mitochondrial disease are no longer considered at increased risk for malignant hyperthermia compared to the general population.

- *Local anesthetics*: Given the multisystem effects of general anesthetics, it is advantageous to use nerve blockade with local anesthetics when possible for surgery. For example, many anesthesiologists routinely block the femoral nerve and provide very light general anesthesia for muscle biopsies from the thigh of young children. Regional blocks can be tailored to provide significant postoperative pain control as well.

  The ability to provide regional analgesia to minimize exposure to general anesthetics is of obvious potential value to patients with mitochondrial defects. However, there are interesting clinical and research reports about the effects of bupivacaine on mitochondrial function, particularly inhibition of acylcarnitine translocase, as well as inhibition of complexes I and III [21, 22]. In general, the clinical ramifications of these findings indicate that strict attention must be given to minimize dosing of local anesthetics in these patients to avoid severe cardiac dysfunction resulting from local anesthetic toxicity.

Postoperative Care

In general, the patient must be very closely monitored for residual respiratory insufficiency, dysphagia and cardiac depression (see Figure 12.1). In addition, all measures to minimize nausea and vomiting must be taken, as the resulting hypovolemia and increased metabolic demand are not well tolerated by mitochondrial patients. Having the patient return to normal feeding as quickly as possible after surgery is very desirable.

Fever and hypothermia/shivering are potentially significant perioperative problems in these vulnerable patients [4, 15]. Anticipation and early treatment of temperature shifts is important, in order to avert the increased metabolic demand of either shivering or temperature elevations. Pain in the postoperative period increases stress and metabolic demand. Appropriate analgesics with use of local anesthetics for postoperative pain relief are required, but the use of acetaminophen is limited due to its reported tendency to oxidize glutathione. Many young patients or patients with developmental delay may not verbalize pain. Opioids carry a risk of respiratory depression and can cause nausea and vomiting, with risk of aspiration, in many patients with mitochondrial disease. Thus, they must be titrated with great care; however, this should not serve as a reason to fail to achieve adequate analgesia.

In the section that follows, we give a few examples of possible management for different sorts of patients with mitochondrial disease.

CASE REPORTS

Case Report 12.1

An eight-month-old male infant presented for a muscle biopsy of the left thigh to diagnose a possible mitochondrial cytopathy. It was estimated that the procedure would take less than a half hour. Past history was remarkable for failure to thrive (patient weighs 5 kg). There was no family history of mitochondrial disease. His mother stated that he was unable to feed effectively. At five months of age, a nasal feeding tube was placed; the patient was dependent on tube feedings since that time. The patient was also hypotonic, unable to sit without support. There was no history of seizures or of cardiac symptoms. No mutations in mitochondrial DNA were found. Review of systems was otherwise normal. MRI findings showed necrotic lesions in the brainstem and cerebellum. Electro- and echocardiograms were within normal limits. An elevated level of lactate was noted on laboratory tests. A tentative diagnosis of Leigh syndrome was made.

Self-Assessment Questions

- The patient presents with a myopathy caused by an unknown defect. What is your anesthetic plan for the patient?
- Would you consider a regional anesthetic? If planning a regional anesthetic as a supplement, which local anesthetics would you consider?

- Would you prefer a volatile agent or an IV anesthetic for maintenance? Which ones?
- How long preoperatively should the patient be fasted? Should the patient be admitted for preoperative IV fluids? What is your preferred maintenance IV fluid?
- Why is a mitochondrial patient at increased risk of perioperative hypothermia?

Discussion

1. It is not unusual for a pediatric anesthesiologist to be presented with the challenge of determining the best management for an infant with undiagnosed myopathy. The differential diagnosis for hypotonia in infancy includes multiple congenital neuromuscular disorders and metabolic disease. Because central core disease, with the associated risk of malignant hyperthermia (MH), is among the most common causes of congenital myopathy, the undiagnosed hypotonic infant presents an anesthetic dilemma. In general, one avoids triggering agents (volatile anesthetics and succinylcholine) in patients at risk for MH; propofol infusions are commonly substituted as the anesthetic in those cases. However, if a mitochondrial defect is suspected, propofol infusions are usually avoided (although there are no evidence-based studies to support this).

    In the case where a persistent lactic acidosis occurs, one would probably suspect mitochondrial disease. Present evidence indicates that patients with mitochondrial defects are not at increased risk for MH. In the absence of some metabolic component, one would likely err on the side of avoiding MH. If one truly cannot decide, a non-triggering, non-propofol-based anesthetic using dexmedetomidine, remifentanil, ketamine and regional anesthetic techniques may be a reasonable approach.

    Patients with a complex I defect may have increased sensitivity to volatile anesthetics. The number of patients studied was small but very significant differences were found. As a result, some additional measurement of anesthetic depth may be useful in these patients.

    A regional anesthetic is often a very useful addition for these patients. The analgesia alleviates postoperative pain, resulting in a more stress-free procedure. In addition, since some patients may be hypersensitive to anesthetics, a regional anesthetic allows for decreases in the necessary general anesthetic to maintain patient comfort. One report does show that the local anesthetic bupivacaine has mitochondrial effects, so it may be wise to use other local anesthetics if possible. In addition, a field block is probably not a good choice due to the possibility of local anesthetic being present in the biopsied tissue, possibly altering the biochemical measurements.

2. It is important to avoid circumstances that place a metabolic burden on these patients. These circumstances include prolonged fasting, hypoglycemia, postoperative nausea and vomiting, hypothermia (with resulting shivering), prolonged tourniquet placement, acidosis and hypovolemia. Hypovolemia presents a special risk in these patients who may also lack the cardiac reserve to increase output in compensation. As a result, most institutions either limit their preoperative fasting time or place preoperative IV lines to give fluids prior to surgery.

    The precise fluids these patients should receive are somewhat variable. In general, it is wise to avoid hypoglycemia in mitochondrial patients, as they may not be able to easily mobilize other substrates for metabolism. However, there are patients who are kept on ketogenic diets to avoid seizures (see Case Report 12.2). Finally, some of these patients are unable to metabolize lactate; thus, it is best to avoid the addition of lactate to their fluids [17]. This may be a more theoretical than absolute consideration, however. The amount of lactate added under normal perioperative conditions should not be enough to significantly change the serum lactate level.

3. Unfortunately, essentially every general anesthetic studied has been shown to depress mitochondrial function. The most notable of these are the volatile anesthetics and propofol, both of which can depress mitochondrial activity at clinical doses. This can lead to decreased metabolic rate with resulting decreases in temperature. This is a double-edged sword since shivering represents a metabolic challenge not well tolerated by these patients. Thus, the anesthetist needs to monitor body temperature closely.

Approach

The patient's mother was encouraged to feed clear liquids containing glucose until two hours prior to surgery, and then oral intake was stopped. The case was scheduled for an 0800 start. Anesthesia was induced with sevoflurane, a volatile anesthetic that is rapidly eliminated through the lungs, by mask ventilation. A BIS© monitor, an integrated EEG that is a surrogate for levels of consciousness, was in place to determine the response to the volatile anesthetic. The patient abruptly lost consciousness and the BIS© decreased to 40 at a sevoflurane end tidal concentration of 0.9 percent, less than half the usual requirement. A peripheral intravenous line was started and the patient was given D2.5 1/2NS at maintenance rates. A point of care serum glucose was 105 mg/dl. Since oral intake was restricted for only two hours, no additional fluid was given to replace volume depletion. The patient was given fentanyl (1mcg/kg) and a single dose of propofol (1mg/kg) to facilitate tracheal intubation. A femoral nerve block was placed while under general anesthesia, using ropivacaine to help provide analgesia for the muscle biopsy. The remainder of the case was done using sevoflurane (1–1.5 percent), 30 percent O_2 in air. The patient initially breathed spontaneously with tidal volumes of only 20–25mls (5mls/kg) and was therefore placed on controlled respiration with tidal volumes of 50–60 mls (10–11mls/kg) to avoid atelectasis of dependent alveoli. The case proceeded without problems and the patient was extubated awake in the operating room within five minutes of the end of the surgical procedure. The patient showed no signs of pain in the recovery room. His parents were advised to give ibuprofen or small doses of opioid as needed for discomfort at the biopsy site. The patient was discharged to home two hours after the end of surgery. Results of the biopsy were returned two weeks after surgery and were consistent with a complex I defect. Mitochondrial complex I dependent respiration rates were only 20 percent of normal.

Case Report 12.2

A seven-year-old with increasing dysphagia and muscle weakness with lagging developmental milestones presented for placement of a permanent feeding gastrostomy. The medical history was remarkable for a diagnosis of MELAS syndrome, with increasing muscle weakness. The patient also had had a stroke at five years of age, which decreased his ability to walk due to residual weakness in the right leg. The diagnosis of mitochondrial disease was originally made on clinical grounds with no biopsy performed. However, subsequent gene sequencing of peripheral blood was notable for a missense mutation in *MTND1*, consistent with the original diagnosis.

During the previous year, the patient had become increasingly unable to swallow and was dependent on a nasal feeding tube for all caloric intake. His weight decreased and he was in the 15th percentile for weight (20.2 kg). In addition, he had become increasingly dependent on suctioning to clear oral secretions.

Self-Assessment Questions

- How long should the patient be fasted preoperatively?
- Should the patient be admitted for preoperative IV fluids?

Discussion

It is important to remember that many of the usual considerations for perioperative patients still are pertinent in mitochondrial patients. The necessity of precautions to avoid aspiration is one such consideration. This case demonstrates that not all patients can be fasted for only two hours; thus adjustments in the fluids need to be addressed. Preoperative admission for IV fluids should always be considered.

Approach

Due to the decreased gastrointestinal motility, the surgical service was uncomfortable with a shortened period of preoperative oral intake. The patient was admitted the night before surgery for placement of intravenous access to give fluids (D5 1/2NS). He was fasted for eight hours (nothing by mouth, nothing in the feeding tube) prior to surgery. Fluids were given at his maintenance rate. The patient was the first case of the day, starting at 0730. Due to the inability to clear oral secretions and the decreased gastrointestinal motility, a rapid sequence intravenous induction was performed. A single dose of the intravenous anesthetic propofol (2 mg/kg) with 1mcg/kg fentanyl, an opioid, and a small dose of rocuronium (0.3 mg/kg), a muscle relaxant, were used for induction. Intubation was performed with cricoid pressure and a cuffed 5.0 oral endotracheal tube was inserted without problems. To maintain anesthesia, the patient was started on sevoflurane. A BIS© monitor was placed and the patient responded normally to the volatile anesthetic. General anesthesia and controlled ventilation were maintained with 3 percent sevoflurane and 30 percent oxygen. To alleviate postoperative pain and avoid opioids that might cause respiratory depression, a transversus abdominis plane block was placed after intubation using 15 ml of 0.2 percent ropivacaine, a block expected to deliver six to eight hours of pain relief. The remainder of the case, which lasted 45 minutes, was uneventful. The muscle relaxation was reversed with small doses of glycopyrrolate and neostigmine. Spontaneous ventilations resumed and the patient was extubated awake in the operating room 10 minutes after surgery was completed. He reported no pain at the site of his incision. The patient was discharged to home two hours after surgery. Postoperative pain was controlled with ibuprofen.

Case Report 12.3

A 13-year-old with progressive kyphoscoliosis and respiratory insufficiency with a known diagnosis of a complex III deficiency (complex III rates 10 percent of normal) was scheduled for placement of Harrington rods. Her previous admission to the hospital for release of flexion contractures of her lower extremities was complicated by a prolonged recovery and mechanical ventilation. Her parents were anxious to avoid a repeat perioperative course.

Preoperative: The patient was known to have restrictive lung disease, with a total lung capacity at 40 percent of predicted. She suffered from seizures that were moderately well controlled by tegretol and a ketogenic diet. She was anemic (Hgb 9.4), thought to be due to her chronic disease and nutritional inadequacies. Cardiac function was normal as was her central nervous system development. She was in the seventh grade and considered the best student in her class.

Self-Assessment Questions

Based on operative length and complexity, as well as patient diagnosis, are the anesthetic considerations for this patient different than those in the prior case?

- IV fluids
- Anesthetics
- Perioperative considerations

Discussion

Patients on a ketogenic diet should *not* have added glucose in their IV fluids. It has also been noticed that many mitochondrial patients on ketogenic diets cannot tolerate even normal amounts of glucose and may become hyperglycemic when glucose administration is higher than usual maintenance. This abnormally sensitive response may occur even if the glucose comes from other sources such as cardioplegia or medications in glucose or related compounds. A prudent approach would be to discuss glucose status with the primary physician to establish limits for glucose exposure. If that is not possible, one must avoid glucose in those patients on a ketogenic diet. It is still best to avoid lactate-containing fluids, i.e. Ringer's lactate. In these patients, a normal or half-normal sodium chloride solution is often preferred. Avoidance of prolonged fasting and maintaining adequate hydration and appropriate glucose replacement during the perioperative period is necessary for patients with mitochondrial disease in order to prevent an acute lactic acidosis and exacerbation.

As noted earlier, it has been reported that patients with a complex I defect may have increased sensitivity to volatile anesthetics. However, the number of patients studied was small and it is too early to say that only complex I patients can be hypersensitive to volatile anesthetics or any other agent. As a result, some additional measurement of anesthetic depth may be useful in these patients. Use of some measure of the processed EEG has been used with success.

Approach

The surgery was scheduled to start at noon, and predicted to last for six hours. Significant blood loss was anticipated during surgery; she was typed and crossed for four units of red cells, and both platelets and fresh frozen plasma were ordered preoperatively. In addition, a cell saver was made available to minimize blood loss. Since both red cells and the cell saver primer fluid normally contain glucose, they were prepared preoperatively especially for this patient without glucose, in order to continue her ketogenic state. The patient was started on intravenous hydration with 1/2NS upon admission at 0800 in the morning, four hours prior to scheduled start of surgery, and her oral intake of clear liquids was stopped at that time.

Intraoperative: A single dose of the intravenous anesthetic propofol (2 mg/kg) with 1mcg/kg fentanyl, an opioid, and a small dose of rocuronium (0.3 mg/kg), a muscle relaxant, were used for induction and placement of the endotracheal tube. Since she was receiving spinal cord monitoring, volatile anesthetics could not be used for maintenance of her anesthetic. In addition, a propofol infusion was felt to be undesirable due to its multiple inhibitory effects on the mitochondrion. She was maintained under general anesthesia with a combination of ketamine, dexmedetomidine, versed and remifentanil. In addition, subarachnoid morphine was given to ensure adequate analgesia both intraoperatively and postoperatively.

Once in the prone position, she was ventilated using relatively small tidal volumes and a high-frequency respiratory rate. Spinal cord monitoring and EEG measurements ensured that adequate anesthetic depth was maintained. The remainder of the case went smoothly and she required only two units of glucose-free red blood cells. She was extubated at the end of the case and awoke rapidly with minimal discomfort. She was transferred to the PICU overnight for observation and then to the general surgical ward the following day.

Summary

Patients with mitochondrial disease require meticulous care of their lungs, since they are at increased risk of respiratory failure from multiple complications of their disease. Surgical procedures require anesthesia, which must provide adequate analgesia but seek to avoid central inhibition of respiratory drive. This often indicates aggressive use of regional analgesia as either an adjunct or an alternative to general anesthesia. In addition, care must be taken to avoid muscle relaxation so that respiratory strength is adequate. Finally, analgesia must be sufficient to ensure that pain does not limit respiratory effort. With these caveats in mind, and appropriate evaluation and planning on the part of multiple caregivers, mitochondrial patients can undergo surgical procedures safely.

References

1. Haas RH, Parikh S, Falk MJ, et al. The in-depth evaluation of suspected mitochondrial disease. *Mol Genet Metab* 2008;**94**(1):16–37.

2. Niezgoda J, Morgan PG. Anaesthetic considerations in patients with mitochondrial defects. *Paediatr Anaesth* 2013;**23**(9):785–793.

3. Ajit Bolar N, Vanlander AV, Wilbrecht C, et al. Mutation of the iron-sulfur cluster assembly gene IBA57 causes severe myopathy and encephalopathy. *Hum Mol Genet* 2013;**22**(13):2590–2602.

4. Falk MJ, Sondheimer N. Mitochondrial genetic diseases. *Curr Opin Pediatr* 2010;**22**(6):711–716.

5. Song Y, Pinniger GJ, Bakker AJ, et al. Lipopolysaccharide-induced weakness in the preterm diaphragm is associated with mitochondrial electron transport chain dysfunction and oxidative stress. *PLoS One* 2013;**8**(9):e73457.

6. Amornvit J, Pasutharnchat N, Pachinburavan M, et al. Fulminant respiratory muscle paralysis, an expanding clinical spectrum of mitochondrial A3243 G tRNALeu mutation. *J Med Assoc Thai* 2014;**97**(4):467–472.

7. Wilnai Y, Seaver LH, Enns GM. Atypical amyoplasia congenita in an infant with Leigh syndrome: A mitochondrial cause of severe contractures? *Am J Med Genet A* 2012;**158A**(9):2353–2357.

8. Barclay AR, Sholler G, Christodolou J, et al. Pulmonary hypertension – a new manifestation of mitochondrial disease. *J Inherit Metab Dis* 2005;**28**(6):1081–1089.

9. Navarro-Sastre A, Tort F, Stehling O, et al. A fatal mitochondrial disease is associated with defective NFU1 function in the maturation of a subset of mitochondrial Fe-S proteins. *Am J Hum Genet* 2011;**89**(5):656–667.

10. Stettner GM, Viscomi C, Zeviani M, et al. Hypoxic and hypercapnic challenges unveil respiratory vulnerability of Surf1 knockout mice, an animal model of Leigh syndrome. *Mitochondrion* 2011;**11**(3):413–420.

11. Quintana A, Zanella S, Koch H, et al. Fatal breathing dysfunction in a mouse model of Leigh syndrome. *J Clin Invest* 2012;**122**(7):2359–2368.

12. Ventura F, Rocca G, Gentile R, De Stefano F. Sudden death in Leigh syndrome: An autopsy case. *Am J Forensic Med Pathol* 2012;**33**(3):259–261.

13. Klopstock T, Jaksch M, Gasser T. Age and cause of death in mitochondrial diseases. *Neurology* 1999;**53**(4):855–857.

14. Driessen JJ. Neuromuscular and mitochondrial disorders: What is relevant to the anaesthesiologist? *Current Opinion in Anaesthesiology* 2008;**21**:350–355.

15. DiMauro S, Schon EA. Mitochondrial respiratory-chain diseases. *N Engl J Med* 2003;**348**(26):2656–2668.

16. Cooper MA, Fox R. Anesthesia for corrective spinal surgery in a patient with Leigh's disease. *Anesth Analg* 2003;**97**(5):1539–1541.

17. Pravdic D, Hirata N, Barber L, et al. Complex I and ATP synthase mediate membrane depolarization and matrix acidification by isoflurane in mitochondria. *Eur J Pharmacol* 2012;**690**(1–3):149–157.

18. Kayser EB, Suthammarak W, Morgan PG, Sedensky MM. Isoflurane selectively inhibits distal mitochondrial complex I in Caenorhabditis elegans. *Anesth Analg* 2011;**112**(6):1321–1329.

19. Morgan PG, Hoppel CL, Sedensky MM. Mitochondrial defects and anaesthetic sensitivity. *Anesthesiology* 2002;**96**(5):1268–1270.

20. Kajimoto M, Atkinson DB, Ledee DR, et al. Propofol compared with isoflurane inhibits mitochondrial metabolism in immature swine cerebral cortex. *J Cereb Blood Flow Metab* 2014;**34**(3):514–521.

21. Weinberg GL, Palmer JW, VadeBoncouer TR, et al. Bupivacaine inhibits acylcarnitine exchange in cardiac mitochondria. *Anesthesiology* 2000;**92**(2):523–528.

22. Onyuksel H, Sethi V, Weinberg GL, et al. Bupivacaine, but not lidocaine, disrupts cardiolipin-containing small biomimetic unilamellar liposomes. *Chem Biol Interact* 2007;**169**(3):154–159.

Endocrinology and Diabetes

Mark Walker, Alison J. Heggie, Andrew M. Schaefer and
Grainne S. Gorman

Mitochondrial Diabetes

Diabetes mellitus is the most common endocrine manifestation of mitochondrial disease.
The m.3243A>G *MTTL1* mutation [1] is the mitochondrial (mt) mutation most commonly
associated with diabetes. The prevalence of the m.3243A>G mutation in unselected diabetic
populations varies between 0 percent and 2.8 percent based on a review of larger studies [2],
and in a UK clinic was found to be 0.13 percent [3]. The most common phenotypic
presentation is maternally inherited diabetes and deafness (MIDD), but diabetes can be
associated with more severe and wide-ranging phenotypes such as mitochondrial encephalomyopathy, lactic acidosis and stroke-like episodes (MELAS).

Several other mtDNA mutations are associated with diabetes and include the m.14709
T>C mutation [4, 5], which may account for up to 13 percent of mitochondrial diabetes in
the northeast of England [6]; the m.8296A>G *MTTK* gene mutation identified in 0.9 percent
of unrelated Japanese patients with diabetes [7]; and the m.14577 T>C *MTND6* mutation
associated with isolated complex I deficiency [8].

Single, large-scale mtDNA deletions have been associated with diabetes in 11 percent of
well-defined clinical cohorts of patients with chronic progressive external ophthalmoplegia
(CPEO) and Kearns-Sayre syndrome (KSS) [6].

Insulin secretion from the pancreatic beta-cells is highly energy dependent and
mitochondrial adenosine triphosphate (ATP) generation is a crucial step in the insulin
secretory pathway (Figure 13.1). Depletion of mtDNA directly impairs mitochondrial
function and results in a decrease in glucose-stimulated insulin secretion [9]. It is no
surprise, therefore, that impaired insulin secretion and diabetes are linked to mitochondrial disease.

The classic clinical features of MIDD are diabetes presenting in early middle age, which
is generally preceded by sensorineural deafness. Most patients will have a normal or low
BMI, and so the key question is whether the patient might have autoimmune type 1 diabetes.
This usually presents with weight loss, the presence of ketones, a low or absent C-peptide
and GAD antibody positivity indicating insulin deficiency and an autoimmune predisposition, respectively (see Case Report 13.1). However, a small proportion of patients with
mitochondrial diabetes can present with clinical and biochemical features of insulin deficiency, and they need to start insulin therapy immediately. The majority of MIDD patients
presents as nonobese type 2 diabetes, and guidelines have been developed for the management of these patients (see cases and [10]). However, these patients are at high risk of
progression to insulin-dependent diabetes, often within the first two to four years after
diabetes diagnosis [2, 6].

Insulin secretion from the beta-cell

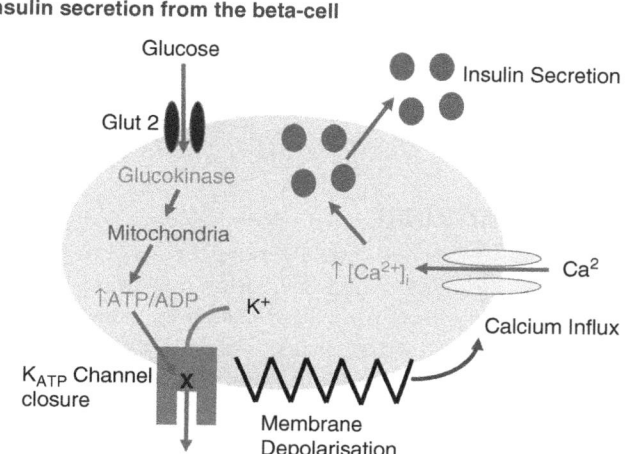

Figure 13.1 Schematic diagram depicting the key steps in glucose-stimulated insulin secretion from the pancreatic beta-cell. Glucose enters the beta-cell via the GLUT 2 transporter, and is then metabolized by the mitochondria to generate ATP. An increase in the ATP/ADP ratio leads to closure of the K_{ATP} channel, cell membrane depolarization and opening of calcium channels. Influx of calcium into the beta-cell stimulates insulin release. (A black and white version of this figure will appear in some formats. For the color version, please refer to the plate section.)

A maternal history alone of diabetes is not a reliable predictor of mitochondrial diabetes, in part because a history of diabetes in one or other parent is common in type 2 diabetes [4]. However, a maternal history of diabetes plus other features of mitochondrial diabetes such as early onset deafness and/or neuromuscular disease should prompt further investigation for mitochondrial diabetes.

Endocrine Disorders

As recently reviewed [2], mitochondrial mutations have been associated with other endocrine disorders. Hypoparathyroidism has been reported in cases of sporadic KSS, and results in the biochemical changes of hypocalcemia, hypomagnesemia and a low circulating parathyroid hormone concentration. Severe hypocalcemia can result in parasthesia and impaired muscle function, and is important to detect as it is easy to manage with replacement therapy. Hypoparathyroidism can enhance bone mass with time, but it's not clear whether this in turn alters the bone structure and propensity to fracture.

Mitochondrial mutations have been associated with abnormalities of the hypothalamic–pituitary axis, and tend to occur in more severe cases of mitochondrial disease, usually MELAS and KSS. Growth hormone deficiency is well described and can contribute to short stature, while gonadotrophin deficiency can lead to impaired gonadal function. This in turn can lead to hormone deficiency in adulthood, which is important to detect and correct in order to preserve bone density and promote general well-being.

Hyponatremia is a common feature of patients with the m.3243A>G *MTTL1* mutation. It can result from a number of interrelated factors, including mitochondrial renal tubular dysfunction and SIADH (syndrome of inappropriate ADH secretion). MELAS patients seem to be susceptible to SIADH, exacerbated by certain drugs such as carbamazepine. Another rare but important potential contributor to hyponatremia is adrenal insufficiency. This has been described in association with KSS, and can be an early feature presenting in childhood. It is usually due to primary adrenal failure and can be quickly diagnosed by short synacthen test. It is crucial to identify and manage with steroid replacement, and to provide guidance on steroid dose adjustment during times of acute clinical/metabolic stress.

Thyroid dysfunction does not seem to be a specific problem in mitochondrial disorders, with the frequency of thyroid disease seemingly comparable to that of the background population.

Case Reports

Case Report 13.1: Clinical Presentation

A 37-year-old lady presented to Medical Admissions with deteriorating diabetes control. She had been prescribed a course of oral steroids for an exacerbation of asthma. She was reviewed by the on-call diabetes registrar, who ascertained that she had been diagnosed with diabetes 12 months earlier. She had been prescribed gliclazide 80 mg bd by her general practitioner, but had to stop because of erratic and often low blood glucose values. At the time of presentation she was managing her diabetes through dietary measures alone. Clinical examination was unremarkable apart from the fact that she was wearing bilateral hearing aids. She was of normal weight.

Abnormal results at the time of admission were: random blood glucose was markedly raised at 28 mmol/l (504 mg/dl) and moderate ketonuria on dipstick. Urea and electrolytes and venous blood gases were all normal. The registrar started insulin therapy and arranged for urgent review in the diabetes clinic.

Self-Assessment Questions

- What further clinical information would you seek?
- What further investigations would you arrange when you see her at the diabetes clinic?
- Why had she encountered difficulties with use of gliclazide?

Answers

- The combination of bilateral hearing loss preceding the development of diabetes in early middle age should raise the possibility of mitochondrial diabetes. The registrar made the connection and went on to ask about the family history. It transpired that both her mother and maternal grandmother had diabetes, and her mother had hearing difficulties requiring the use of a hearing aid. Taken together, this information points to a diagnosis of maternally inherited diabetes and deafness (MIDD).
- At the diabetes follow-up clinic, DNA was sent for mitochondrial mutation testing and showed that she carried the m.3243A>G *MTTL1* mtDNA mutation, confirming the diagnosis of MIDD. In view of the history of shortness of breath, an ECG and echocardiogram were requested to explore whether cardiac dysfunction was contributing to her symptoms. Although she had been managing her diabetes through lifestyle measures alone, the acute asthma attack and steroid therapy resulted in a very high random blood glucose and ketonuria pointing to insulin deficiency under conditions of stress. To assess this further, GAD and islet cell antibodies were checked and returned normal, making autoimmune type 1 diabetes unlikely. To assess her residual insulin secretory reserve, a random C-peptide was checked. C-peptide is co-secreted with insulin and serves as a marker of endogenous insulin secretion. Her C-peptide returned at 0.36 (normal range 0.34–1.8 nmol/l), indicating that she had some residual insulin secretion – but clearly not enough to control her blood glucose

levels at times of acute illness. In view of these findings, and the future risk of further episodes of asthma relapse and steroid therapy, the decision was taken to continue insulin therapy long term.

- Gliclazide is a sulphonylurea and patients with MIDD can show increased sensitivity to this class of agent, leading to an increased risk of recurrent hypoglycemia.
  If sulphonylureas are to be prescribed to a patient with mitochondrial diabetes, then it is advised to start a low dose (for example, 40 mg od) and introduce home blood glucose monitoring.

Case Report 13.2: Risk Factor Management

On a routine general practice health check, a 52-year-old man was found to have newly diagnosed type 2 diabetes with an HbA1c of 68 mmol/mol. There was no personal or family history of coronary vascular disease (CVD) and he was a nonsmoker. However, he was hypertensive on repeat testing and prescribed an ACE inhibitor. His initial blood lipid profile had shown a raised total cholesterol 6.2 mmol/l, HDL cholesterol 0.9 mmol/l and non-fasting triglycerides of 2.3 mmol/l. He was moderately overweight with a BMI of 28.4 kg/m [2]. Following current NICE guidance, his general practitioner (GP) had used QRISK2 to determine that his 10-year CVD risk was 24 percent and had accordingly prescribed atorvastatin 20 mg OD. He was given lifestyle advice and prescribed metformin, initially 500 mg bd.

Over the next two years, his blood glucose control deteriorated despite maximal doses of metformin and gliclazide. His GP referred him to the local diabetes service for further management.

He was seen by the registrar, who noted his history, but also that the patient had used hearing aids since early childhood. There was a maternal history of diabetes, deafness and dementia, presumed Alzheimer's disease. He tried to walk three miles per day, but had recently struggled because of proximal muscle pains. His clinic bloods showed a raised HbA1c of 77 mmol/mol, a total cholesterol of 3.4 mmol/L and normal triglycerides, and renal function was normal. His blood pressure (BP) was 128/72, and there was no evidence of small vessel complications.

The registrar queried the diagnosis of MIDD and was later confirmed with the identification of the m.3243A>G mt DNA mutation. In view of this diagnosis, the patient was asked to stop the metformin and switched to a DPP4 inhibitor together with the gliclazide. However, at three-month follow-up, the HbA1c had increased to 87 mmol/mol and the patient complained of increasing muscle pains and weight gain.

Self-Assessment

- How would you manage his blood glucose?
- How would you investigate and manage the muscle pains?
- What is his 10-year CVD risk target?

Answers

- Metformin is widely used and effective in type 2 diabetes. Its mode of action is not completely understood, but it has been shown to inhibit complex 1 activity *in vitro*. For this reason, the general recommendation is to avoid it in mitochondrial diabetes.

However, it is often well tolerated and is clinically effective in MIDD as in this patient. His glycemic control deteriorated following the change of therapy, and indeed he asked to restart the metformin at his follow-up appointment. This was agreed, but as a precaution his lactate levels were periodically checked after a short exercise test.

- DPP4 inhibitors increase circulating GLP-1 levels and promote insulin secretion. They must not be used in patients who have a history of and/or increased risk of pancreatitis.
- Many patients with MIDD have gastrointestinal complications, in particular constipation. These problems can be exacerbated by metformin and other diabetes treatments, and so may hamper the options for diabetes management. In this patient, his blood glucose control remained suboptimal despite triple therapy (metformin, DDP4 inhibitor and gliclazide), and so he was started on insulin (in place of the gliclazide and DPP4 inhibitor) and continued the metformin.
- Despite an improvement in his blood glucose control, he continued to be troubled by muscle pains. A creatine kinase was normal, but it was decided to see if his symptoms were related to the statin. The atorvastatin was stopped for three months, and this improved his leg discomfort. However, his lipid profile relapsed with a total cholesterol of 6.4 mmol/L and HDL cholesterol of 1.0 mmol/L. It was agreed to reintroduce low-dose pravastatin with review of creatine kinase levels. He tolerated pravastatin 20 mg od, and this improved his lipid profile so that his total to HDL cholesterol ratio was acceptable at 4.0.
- According to UK national guidance, his 10-year cardiovascular disease risk should be less than 10 percent. This was achieved in the present patient through careful control of his BP with an ACE inhibitor and the titration of pravastatin to 20 mg od. However, there is a clear degree of pragmatism, especially for patients with MIDD, in balancing potential benefits against the side effects of commonly used drugs such as the statins.

Case Report 13.3: Diabetes and Pregnancy

A 33-year-old lady with a family history of MELAS was tested for diabetes in the first trimester of her second pregnancy following a period of weight gain, and found to have a slightly elevated random blood glucose of 14 mmol/l (252 mg/dl) and an HbA1c of 54 mmol/mol (normal <42 mmol/mol). She also had a background of hypertension, currently controlled on methyldopa (BP 124/70). She had no osmotic symptoms and there was no ketonemia (indicative of insulin deficiency). She had had a normal two-hour 75 g oral glucose tolerance test (OGTT) at age 21 and normal HbA1c tests on subsequent monitoring, and she did not develop gestational diabetes in her previous pregnancy. She had therefore never had screening for any diabetes-related complications, but renal function and foot examination were normal.

She was known to have inherited the m.3243A>G mutation from her mother, who had several features including stroke-like episodes; the patient's degree of mtDNA heteroplasmy level was 50–55 percent by urine cellular analysis. At age 25, she was found to have elevated liver function tests that led to a diagnosis of nonalcoholic fatty liver disease and insulin resistance. She had experienced some difficulties with hearing, particularly in noisy environments, and audiometry revealed mild sensorineural deafness. Lactate was intermittently raised and she also suffered myalgia on exertion.

Self-Assessment

- How would you determine whether this lady has gestational diabetes mellitus (GDM)?
- What treatments are considered safe for treating diabetes in pregnancy? Which would be most appropriate here?
- What is the significance of the HbA1c? What does it suggest about management after delivery?
- Is there anything else to consider in this patient's management?

Answers and Discussion

- Pregnancy can cause or exacerbate glucose intolerance, mainly through insulin resistance. This leads to GDM in 4–5 percent of pregnant women in the UK, even with no previous history of diabetes during pregnancy [11]. It tends to develop during the second or third trimesters. Poorly controlled diabetes during pregnancy increases the likelihood of both maternal and fetal complications such as preeclampsia, macrosomia, premature birth, neonatal hypoglycemia and jaundice [12].
- Women with mitochondrial disease have a higher risk of GDM, particularly those with certain mutations (m.3243A>G and 14709 T>C), and those with a strong family history of diabetes and associated mitochondrial disease. Women with any of these conditions should have an OGTT at 20 weeks' gestation. Women with a previous history of GDM should be offered self-monitoring of blood glucose and/or an OGTT at time of booking, followed by a further OGTT at 24–28 weeks if the initial OGTT is normal. The current limits used for diagnosis of GDM are tighter than those for non-gestational diabetes; on OGTT the fasting glucose limit is 5.6 mmol/l (100 mg/dl) and the two-hour postprandial limit 7.8 mmol/l(140 mg/dl).
- The treatments generally used in GDM are metformin and insulin, along with glibenclamide in certain circumstances. The target blood sugars once GDM is diagnosed are: fasting 5.3 mmol/l (95 mg/dl); one hour postprandial 7.8 (140 mg/dl) mmol/l; and two hours postprandial 6.4 mmol/l (115 mg/dl) [11]. Metformin is best avoided in mitochondrial diabetes given the theoretical risk of elevating lactate further and sulphonylureas can, as discussed in case study 13.1, lead to problematic hypoglycemia. Given this and the degree of elevation of subsequent blood glucose levels, she agreed to start insulin as a basal bolus regime in order to achieve rapid glycemic control and to reduce her risk of complications. She would also require early retinal screening, as there is increased risk of deterioration in any retinopathy both in pregnancy and following a rapid improvement in glycemic control. She should be monitored particularly carefully for hypertension and proteinuria, as there is preliminary evidence that patients with mitochondrial disease are at increased risk of preeclampsia. Aspirin should be used from 12 weeks' gestation until delivery if there is a history of preeclampsia [13].
- The glucose intolerance displayed in GDM resolves rapidly after delivery and even women on insulin can safely stop their treatment at this point [11]. With this patient, her HbA1c was elevated at presentation (which reflects glycaemia over the preceding two to three months), and she was found to have diabetes relatively early in pregnancy. These findings suggest that she likely had underlying diabetes before pregnancy and is more likely to need ongoing treatment afterward. Postnatally her blood glucose levels remained elevated and her 12-week HbA1c check was still above normal at 50 mmol/

mol in keeping with persistent diabetes. This was initially addressed with dietary change and weight loss. Her HbA1c gradually rose, however, and when she had finished breastfeeding, she was switched to a DDP4 inhibitor.
- Finally, consider early discussion about genetic screening of the child with the family and seek expert involvement.

Case Report 13.4: Diabetes and Renal Failure

A 48-year-old lady was referred to the joint pancreas and kidney transplant assessment clinic. She had been treated for hypertension since her early 30s and developed diabetes aged 39. She was initially treated by her local hospital with tablets, but was quickly switched to insulin with a clinical diagnosis of late-onset autoimmune type 1 diabetes. Her blood glucose control was suboptimal and her renal function started to fail, which was presumed to be a consequence of her hypertension and diabetes. She had no other complications of diabetes. At the age of 45, she required hemodialysis and a year later underwent living donor kidney transplantation. Her general health improved, but her diabetes continued to be difficult to control, and so she was referred for consideration of pancreas after kidney transplantation.

During the review, it transpired that she had used hearing aids since her early 20s, but had hearing impairment of the left ear since birth. Her four siblings had all developed insulin-treated diabetes in their 40s, but none had known deafness. Her mother developed diabetes aged 50 and had required hearing aids in her 60s. The patient was worked up and underwent successful pancreas transplantation at the age of 50. Insulin therapy was stopped immediately post-transplant and her blood glucose levels have remained normal three years out from the transplant.

Self-Assessment

- What other investigations would you consider to determine the cause of this lady's diabetes?
- What are the potential causes of this lady's initial renal failure?
- What are the benefits of pancreas transplantation in monogenic forms of diabetes?

Answers

- A number of elements to this lady's presentation are atypical. The diagnosis of late-onset type 1 diabetes had been reached on clinical grounds. However, GAD antibody titer was negative and C-peptide level was weakly positive when checked by the transplant clinic. These would argue against autoimmune diabetes. The strong family history of diabetes raises the possibility of monogenic forms of diabetes. While there was a history of deafness in the patient, it was slightly unusual as she had impairment in the left ear from birth. Furthermore, her mother appeared to develop hearing impairment after developing diabetes, and none of her diabetic siblings had known hearing impairment. On this basis, the initial investigations focused on whether she might have maturity-onset diabetes of the young, especially as mutations in the *HNF1B* and *HNF1A* genes have been linked with diabetes and renal failure. However, mutation screening was negative. She was then screened for mitochondrial diabetes and found to carry the m.3243A>G mtDNA mutation.

- This case illustrates that clinically apparent hearing deficit does not always precede the development of diabetes in MIDD, but that the combination of maternally inherited diabetes plus evidence of hearing impairment in one or more family members should raise the suspicion of the potential diagnosis.

- The renal failure had been assumed to be due to a combination of hypertension and diabetes. However, the diabetes was diagnosed just six years before she required dialysis for end-stage renal failure, and her BP had been well controlled. It is unusual for diabetes to cause such rapid deterioration in renal function, and it is very unusual for diabetic nephropathy to develop in the complete absence of diabetic eye disease. The most likely cause of the renal dysfunction is glomerulosclerosis related to the mitochondrial disease (see Chapter 15).

- Pancreas transplantation is primarily indicated for patients with type 1 diabetes who have developed renal failure and need a kidney transplant too. Under these circumstances, the kidney and pancreas are retrieved from a nonliving donor and transplanted simultaneously. However, it is recognized that patients with non-autoimmune monogenic diabetes can benefit from transplantation. The aim is to normalize blood glucose levels in order to try to avoid future diabetes complications. This is especially relevant to monogenic diabetes and MIDD when the diabetes usually presents in early adulthood, allowing plenty of time for complications to develop in subsequent years. Furthermore, normalizing the blood glucose levels helps to protect the kidney transplant from future damage. The present patient had been fortunate to receive a kidney transplant from her spouse and was already established on antirejection therapy. This provided the opportunity for subsequent nonliving pancreas transplantation, which resulted in the cessation of insulin and the normalization of her blood glucose levels. In hindsight, she was fortunate that her spouse had donated the kidney. Donation by a maternal relative could have been problematic as they would have carried the m.3243A>G mutation and the transplanted organ susceptible to early failure. Finally, there is the concern in type 1 autoimmune diabetes patients undergoing transplantation that the ongoing autoimmune process might affect the pancreas transplant and limit its survival. Clearly, this is not an issue in non-autoimmune monogenic diabetes, and so the prospects for long-term graft survival are better.

References

1. Van den Ouweland J, Lemkes H, Ruitenbeek W, et al. Mutation in mitochondrial tRNALeu (UUR) gene in a large pedigree with maternally transmitted type 2 diabetes and deafness. Nat. Genet 1992; 1: 368–371.

2. Schaefer AM, Walker M, Turnbull DM, Taylor RW. Endocrine disorders in mitochondrial disease. Molecular and Cellular Endocrinology 2013; 379: 2–11.

3. Newkirk JE, Taylor RW, Howell N, et al. Maternally inherited diabetes and deafness: Prevalence in a hospital diabetic population. Diabetic Med 1997; 14: 457–460.

4. Choo-Kang A, Lynn S, Taylor G, et al. Defining the importance of mitochondrial gene defects in maternally inherited diabetes by sequencing the entire mitochondrial genome. Diabetes 2002; 51: 2317–2320.

5. McFarland R, Schaefer AM, Gardner JL, et al. Familial myopathy: New insights into the T14709 C mitochondrial tRNA mutation. Ann. Neurol 2004; 55: 478–484.

6. Whittaker, RG, Schaefer AM, McFarland R et al. Prevalence and progression of diabetes

in mitochondrial disease. *Diabetologia* 2007; **50**: 2085–2089.

7. Kameoka K, Isotani H, Tanaka K, et al. Impaired insulin secretion in Japanese diabetic subjects with an A-to-G mutation at nucleotide 8296 of the mitochondrial DNA in tRNALys. *Diabetes Care* 1998; **21**: 2034–2035.

8. Tawata M, Hayashi J, Isobe K, et al. A new mitochondrial DNA mutation at 14577 T/C is probably a major pathogenic factor for maternally inherited Type 2 diabetes. *Diabetes* 2000; **49**: 1269–1272.

9. Nile DL, Brown AE, Kumaheri MA, et al. Age-related mitochondrial DNA depletion and the impact on pancreatic beta cell function. *PLoS One.* 2014; **9**(12) e115433.

10. Wellcome Centre for Mitochondrial Research: Diabetes Guidelines. www .newcastle-mitochondria.com/wp-content/ uploads/192016/03/Diabetic-Guideline .pdf.

11. NICE (2015) Diabetes in pregnancy: Management of diabetes and its complications from preconception to the postnatal period. *NICE Guidelines NG3*.

12. Perkins J, Dunn JP, Jagasia SM. Perspectives in gestational diabetes mellitus: A review of screening, diagnosis and treatment. *Clinical Diabetes* 2007; **25** (2):57–62.

13. Wellcome Centre for Mitochondrial Research: Guideline Development Group (2013) *Newcastle Mitochondrial Disease Guidelines: Pregnancy in Mitochondrial Disease* (2nd). www.newcastle-mitochondria.com/wp-content/uploads/19 2012/09/Pregnancy-Guidelines.pdf.

Gastroenterology and Hepatology

Irenaeus F. M. de Coo and Jessie M. Hulst

General Introduction

Gastrointestinal (GI) complaints are common in general medicine, but it is always difficult to weigh up the meaning and significance of the symptoms, especially in patients with a multisystem complex disorder. In many mitochondrial diseases, features such as a failure to thrive, loss of appetite and gastroesophageal reflux are common problems affecting up to 15 percent of patients. Complaints may be vague and include reflux, regurgitation, bloating (with or without abdominal pain) and chronic constipation. These symptoms may remain undetected unless specifically requested, and liver dysfunction often remains subclinical until a late stage.

GI involvement in mitochondrial disease is common in patients with cerebral pathology, particularly those with spasticity and extrapyramidal movement disorders. Patients with brainstem pathology that causes stridor and breathing pattern irregularities are also at high risk of GI complications. Patients suffering from seizures may also find that their seizures act as a common trigger of gastric reflux and vice versa.

Severe GI symptoms such as malnutrition and anorexia can also be a primary manifestation of a mitochondrial disease, as observed, for example, in mitochondrial neurogastrointestinal encephalopathy (MNGIE). In the MitoPhenome database (www.mitophenome .org/) [1], 28 distinct GI phenotypes are recognized features of pathogenic mutations of nuclear or mtDNA genes, and 132 mitochondrial disease genes have GI and liver complaints as part of the clinical spectrum. Feeding difficulties in general, vomiting, failure to thrive and liver disease are the most common. Additional, but nonpathogenic mitochondrial DNA variants may also act as disease risk factors for several common GI diseases that are not part of mitochondrial syndromes. For example, possible associations between mtDNA variation and appetite regulation, gastric emptying and abdominal pain have been reported in some centers. While hepatic dysfunction is less common, in both Alpers Huttenlocher disease and POLG-related diseases, liver pathology might be a primary feature of these disorders.

Clinical Presentation

Many patients with mitochondrial disease due to either nuclear DNA or mtDNA mutations have at least some GI symptoms. Luminal pathology in GI symptom (i.e., symptoms affecting the esophagus, stomach and small and large intestines) is most often encountered, with symptoms relating to the liver, biliary tract and pancreas less frequently described and restricted to specific disease syndromes. Mitochondrial disorders only rarely present with isolated GI features, but are frequently seen in combination with other symptoms and signs.

In common mitochondrial disorders such as m.3243A>G MELAS or MELAS-like phenotypes, together with many forms of Leigh disease, patients present early in the disease course with subtle GI symptomatology. This often progresses during follow-up to cause severe GI complaints requiring interventions such as enteral feeding via nasogastric tube or gastrostomy. In a minority, patients go on to develop chronic intestinal pseudo-obstruction (CIPO). This complication can be all too often misinterpreted as an obstructive ileus or, in particular in the oligosymptomatic phase, as a psychogenic problem, raising the possibility of significant mismanagement if not recognized.

CIPO is defined as an intestinal dysmotility disorder characterized by chronic symptoms suggesting bowel obstruction, but in the absence of fixed or occluding lesions. CIPO results from a variety of disorders affecting the enteric neuromuscular system. CIPO can be further classified as neuropathic (inflammatory or degenerative), myopathic or mesenchymopathic. It may affect any portion of the GI tract, although it is most common in the small bowel, causing impaired peristalsis. When the distal esophagus is involved, patients present with symptoms of ineffective peristalsis or even achalasia. Stomach involvement presents with delayed gastric emptying, and colonic involvement, while usually less prominent, can present with a delay in transit time. Anorectal involvement as part of CIPO most commonly presents with a loss of sensory function and a decrease of resting tone and squeeze pressure [2, 3].

Another common mitochondrial disorder where patients often develop GI dysfunction is the single-mtDNA deletion disorder Kearns-Sayre syndrome (KSS). In this patient cohort, dysphagia and swallowing problems are common, though the cause of this is not fully established.

GI symptoms are, as pointed out previously, a common phenomenon in many different mitochondrial disease phenotypes. A rare GI manifestation is hepatic dysfunction. The involvement of the liver appears to be more restricted to certain rare mitochondrial disease genotypes, or occurs as a late development of more common mitochondrial disease in the context of generalized multi-organ involvement. However, it does appear that some genotype-phenotype patterns of GI dysfunction are beginning to emerge in mitochondrial disorders (see later in this chapter). In the next paragraph some examples are given.

Mitochondrial disease syndromes with defects in genes required for mtDNA maintenance and translation (expression) (e.g., *FARS2*, *POLG*) appear more likely to present with hepatic and GI manifestations. Many of these genes influence the replication of mtDNA, causing a mitochondrial depletion syndrome (MDS). The MDS caused by mutations in *BCSL1*, *DGUOK*, *C100RF2*, *MPV17*, *SUCLG1* or *PEO1* commonly leads to hepatopathy [4]. Mutations in some of these genes can cause an early-onset liver disease with growth retardation such as the severe GRACILE syndrome that occurs due to a mutation in *BCS1L* causing growth retardation, aminoaciduria, cholestasis, iron overload, lactic acidosis and early death. Mutations in nuclear-mitochondrial translation factor genes, such as *TRMU* coding for a tRNA modifying enzyme, and *GFM1*, which codes for an elongation factor *mtEFG1*, should also be included in the diagnostic work-up of neonatal liver disease due to their predilection to cause neonatal hepatic dysfunction.

Mitochondrial fatty acid oxidation disorders, among which are the long-chain fatty acid oxidation defects (e.g., very-long-chain acyl-CoA dehydrogenase deficiency [VLCADD], long-chain hydroxyacyl-CoA dehydrogenase deficiency [LCHADD] / mitochondrial trifunctional protein [MTP] deficiency, carnitine palmitoyltransferase 1 [CPT1] and 2 [CPT2]), are a clinically heterogeneous group and may present at any age depending on how the genetic defect influences enzyme activity. A general rule of thumb is that early-onset phenotypes manifest with either hypertrophic cardiomyopathy, hepatic encephalopathy and severe hypo-ketotic hypoglycemia, or a combination of these findings. On the other hand, late-onset forms present with rhabdomyolysis, muscular weakness and exercise intolerance. As metabolic derailment can be prevented by avoidance of fasting and a special high-carbohydrate, low-fat diet, several long-chain fatty acid oxidation defects have been added to neonatal screening programs worldwide.

Clinical Examination

First of all, it is important to assess the patient's nutritional status with attention to height (or body weight for height in infants), weight, body mass index (BMI) and additional anthropometric measurements such as upper arm circumference and skinfold thickness. These parameters should be compared to reference values and expressed as standard deviation scores (or z-scores). Body composition can be assessed using bioelectrical impedance analysis (BIA), dual-energy x-ray absorptiometry (DEXA) or air displacement plethysmography (ADP).

Inspection: During the physical examination, it is important to pay attention to the neuro-logical, respiratory and connective tissue symptoms that may affect swallowing (Table 14.1). Examination should include oral-motor and laryngeal structures. Myopathic facies, nasal speech (palatal weakness), dysphonia (abnormal voice) and dysarthria (abnormal speech articulation) are signs of motor dysfunction. Inspection of the oral cavity for mucosal lesions, dentition and symmetry of the palatal arch during phonation is important to detect mechanical problems that may affect oral intake. The gag reflex is elicited by touching a tongue blade first to one then to the other pillar of the palatal arch. A gag reflex can also be elicited in most normal persons; however, absence of a gag reflex does not necessarily indicate that a patient is unable to swallow safely. If the patient says "AA" and the palate fails to elevate but does elevate during the gag reflex, the patient has to have an upper motor neuron lesion. The pulling of the palate to one side during gag reflex testing indicates lower motor neuron or weakness of the muscles of the contra-lateral palate and suggests the presence of unilateral brainstem pathology.

It is important to observe the patient's swallowing pattern in the act of drinking a few milliliters of water. Drooling, delayed swallow initiation, coughing, throat clearing or a change in voice quality may indicate a problem. Some patients show a delayed cough response. After the swallow act, the patient should be observed for a minute or more to see if there is a delayed cough response. Finally, it is important to inspect the abdomen looking for scars, herniation or distension and dilated veins. Observe together with assessing if the patient is restless or has their knees in a pulled up position as may occur with cases of peritonitis and intussusception.

Abdominal Examination

Auscultation, percussion and palpation: Auscultation for the presence or absence of bowel sounds should be performed. In end-stage mitochondrial or in advanced MNGIE syndrome, percussion of the patient's abdomen can aid in determining if features of peritonitis are present. Painful percussion points to peritonitis. Percussion may also reveal features of obstruction or perforation, together with the estimation of spleen and liver volume, and will assist in detecting the presence of ascites. Percussion should be performed

Table 14.1 Clinical features of gastrointestinal disease in patients with mitochondrial disorders

Sign	Symptom/ Disease
General	
Shrunken face	Cachexia/Malnutrition
Decreased subcutaneous fat	Cachexia/Malnutrition
Sarcopenia	Cachexia/Malnutrition
Peripheral edema	Protein-losing enteropathy
Restless	Intestinal or biliary obstruction
Skin	
Jaundice	Cholestatic liver disease
Excoriation/pruritus	Cholestatic liver disease
Mouth	
Excessive drooling	Esophagus dysphagia Swallowing dysfunction
Abdomen	
Distended abdomen	Constipation Ascites Hepatosplenomegaly Chronic liver insufficiency Pseudo-obstruction syndrome Proximal intestinal obstruction
Protruding umbilicus	Ascites
Bulging flanks	Ascites
Spider naevi	Chronic liver failure (portal hypertension)
	(Para-) umbilical herniation, epigastric herniation, cicatricial hernia
Scars	Status after abdominal surgery, trauma
Inguinal region	
Swelling	Inguinal or femoral hernia
Anal region	
Perianal redness	Chronic diarrhea Carbohydrate malabsorption
Soiling	Constipation Encopresis

supine before being performed laterally, away from the umbilicus, with a shift in tone toward mute indicative of ascites.

Palpation: Palpation of the abdomen has to be performed with warm hands. With superficial palpation note the active and passive muscle tone, pain and palpable resistance. Deeper abdominal palpation shows the presence of any rebound tenderness or any abnormal masses. Finally an organ-specific palpation should be performed to look for hepatospleno-megaly and upon indication, a rectal investigation may be performed.

Clinical Investigations

Conventional abdominal X-ray: A conventional abdominal X-ray investigation can be used to look at the distribution of air and feces in the bowel, the presence of air outside the intestine (sign of perforation), correct placement of feeding tubes, distension of bowel segments, ileus and volvulus.

Abdominal ultrasound: Abdominal ultrasound studies are used to gain information about the size and aspect of the intra-abdominal organs such as the liver, gallbladder, spleen, pancreas, intestine, appendix, kidneys and bladder. Also, the presence and amount of ascites can be evaluated. With Doppler mode the vascularization, e.g., portal circulation can also be investigated.

Magnetic resonance imaging: MRI of the brain with particular attention being paid to the opercular region (opercular syndrome [5]), the basal ganglia, brainstem and the medulla oblongata (all involved in coordination of the swallowing process) is important in cases of suspected neurogenic dysphagia.

Videofluorography: When clinical examination with the SLTP (speech and language thera-pist) has not given enough information as to the cause of dysphagia, the following diagnostic methods can be used to localize the dysphagia in terms of oral, pharyngeal or esophageal pathology and helps to explain the underlying etiology of any dysphagia.

Videofluorographic swallowing study (VFSS) is the first choice to observe the patient's swallow mechanism. It clearly demonstrates the anatomy and the motions of these struc-tures and the passage of the iodinated or barium-contrast bolus through the oral cavity, pharynges and esophagus. If penetration of the opening of the trachea or aspiration into the trachea occurs or food is retained after swallowing, the next step is to evaluate the quantity of retained food, the mechanism of retention or aspiration and the patient's response (e.g., coughing, choking or discomfort).

When a VFSS is not feasible, a fiber-optic endoscopic examination of swallowing (FEES) is helpful. With FEES, a trans-nasal laryngoscope is used to assess pharyngeal swallowing. It can identify penetration/aspiration and pharyngeal retention after the swallow. FEES does not show the motion of the surrounding structures or the food bolus during the swallow.

Manometry: Understanding possible causes of underlying etiology of GI dysmotility can often be obtained from intestinal manometry. Esophageal manometry will reveal abnormalities in esophageal motility in more than 70 percent of patients with CIPO, and antroduodenal manometry may also reveal characteristic motility abnormalities in the small bowel [6, 7].

Biopsy: For detailed investigations in mitochondrial diseases, a tissue biopsy from the most affected organ (e.g., liver) with morphological, biochemical and mtDNA deletion analysis can be helpful in determining the diagnosis (see Chapter 4). A biopsy of the intestine often does not add information to a diagnosis of a mitochondrial disorder (MD), which can often more easily be gained by investigating another tissue such as skeletal muscle. Where CIPO is suspected, extensive morphological investigation of intestinal biopsies, including a full thickness biopsy to investigate the different intestinal layers and the Cajal cells, can be helpful, but obviously necessitates an invasive investigation.

Mitochondrial hepatopathy presents histopathologically with micro- and macro-vesicular steatosis, cholestasis, hepatocellular degeneration and swelling, lobular inflammation, portal fibrosis and, eventually, cirrhosis. In multiple depletion syndromes and Pearson syndrome, hemosiderosis can also be noted.

Laboratory Investigations

Investigations for mitochondrial disease for GI dysfunction:

- Plasma glucose (may detect diabetes mellitus, increasing likelihood of mitochondrial disease)
- Liver, biliary and pancreatic enzymes (ASAT, ALAT, GT, AF, bilirubin – total and direct, albumin and coagulation, amylase and lipase)
- Fasting plasma lactate and pyruvate level
- Plasma and urinary thymidine and deoxyuridine
- Urinary organic acids, amino acids, purine and pyrimidine metabolites
- Plasma acylcarnitines
- Serum albumin, prealbumin, lymphocyte count, C-reactive protein (monitor severity of disease)
- Serum calcium, iron, vitamin B12, folate, A, D, E and K (via prothrombin time/ international normalized ratio), vitamin B1 and nicotinamide (monitor poor intake and malabsorption)

Genetic analysis and biomarkers: Liver and kidney disease might be caused by mitochondrial depletion syndromes, as discussed earlier. We suggest employing the diagnostic paradigm outlined in Chapters 3 and 4 to test for mtDNA point mutations, deletions and depletion in the first instance. If clinical suspicion of a mitochondrial disease that is confined to the GI system, or in which blood, urine or muscle is suspected not to reflect the underlying mitochondrial dysfunction in the GI system, then a biopsy of the appropriate GI tissue should be undertaken. Take into account that the mtDNA heteroplasmy percentage in the different cell layers might differ considerably.

The role of mitochondrial biomarkers in clinical diagnostics is growing. The specific biomarkers that might contribute to a mitochondrial diagnosis are lactate, alanine and FGF-21 and GDF15, but it is important to remember that all of these markers may be abnormal in patients with liver disease not due to a primary mitochondrial disorder.

Managing Complications

Vomiting: For MD-associated cyclic vomiting, there is no proof of efficacy for any specific pharmacotherapy. Among the food supplements and vitamins that have been tried are CoQ_{10}, L-arginine, riboflavin, niacin, L-carnitine and lipoic acid. Supportive treatment, e.g., vitamins for amelioration of the disorder, can be considered, and a low-carbohydrate diet may decrease the lactic acidosis.

Dysphagia: In MELAS and Leigh syndromes, dietary modification, compensatory maneuvers and swallow therapy may be helpful. Swallow therapy, which is a form of rehabilitation, can be divided into three groups: compensatory techniques (i.e., postural maneuvers), indirect therapy (exercises to strengthen swallowing muscles) and direct therapy (exercises to perform while swallowing). Maintaining oral feeding often requires compensatory techniques to reduce aspiration or improve pharyngeal clearance. Advice given by the speech and language therapist in terms of adjusting posture, consistency of fluids or techniques may reduce the swallowing problems.

Gastroesophageal reflux and strictures: In all but the most straightforward cases, input from gastroenterologists, internal medicine physician or a pediatrician is recommended. For patients with gastroesophageal reflux disease, recommendations include dietary modification, cessation of eating before bedtime, remaining upright after eating, pharmacologic therapy (proton pump inhibitors and and/or prokinetics can be tried) and smoking cessation. In case of peptic stricture of the esophagus, achalasia of the lower esophageal sphincter or diffuse esophageal spasm, the usual approach is to start with pharmacologic therapy.

Liver failure: In some cases of liver failure due to mitochondrial disease, a liver transplantation has been performed. However, if liver failure is successfully treated by transplantation, even in an oligosymptomatic patient, other organ manifestations associated with the patient's mitochondrial disorder might appear during follow-up, and long-term follow-up data are lacking.

Insufficient feeding/malnutrition: Enteral mechanical feeding (via nasogastric tube [NG]) may be necessary to bypass the oral cavity and pharynx in cases with severe dysphagia. In general, NG feeding is indicated in any patient who is unable to achieve adequate alimentation and hydration by oral intake. (Figure 14.1)

In patients with aspiration, NG feeding is not always required because, with a combination of dietary modifications and the use of compensatory maneuvers, most patients with mild forms of aspiration can learn to take sufficient food and drink by mouth to meet nutritional requirements. Patients with an impaired level of consciousness, severe aspiration, silent aspiration, esophageal obstruction or recurrent respiratory infections often do require tube feeding to supply enteral nutrition (EN).

A percutaneous endoscopic gastrostomy (PEG) (or surgical gastrostomy) is commonly used for administration of long-term EN. However, this approach is itself associated with an increased risk of gastroesophageal reflux, which can lead to aspiration pneumonia. In case of recurrent life-threatening aspiration pneumonia, a fundoplication procedure of the distal esophagus may be indicated after discussion with GI, medical and surgical teams. In case of CIPO with involvement of the stomach or other forms of gastric emptying problems, feeding through a jejunostomy may be necessary. A jejunostomy feeding tube is surgically placed in the small intestine (jejunum). Both the gastrostomy and jejunostomy can act as an outlet if needed to decrease pressure and pain in the bowel.

When EN via gastrostomy or jejunostomy prove ineffective, parenteral nutrition (PN) may be necessary to maintain a good nutritional status and to relieve GI complaints.

Long-term (home) PN is usually the last step after a trial of different types and modes of delivery (bolus versus continuous) of EN, a trial of medication (prokinetics) and a trial of post-pyloric feeding (in case of gastric emptying problems). Children who need (home) PN should be followed up by a specialized intestinal failure team.

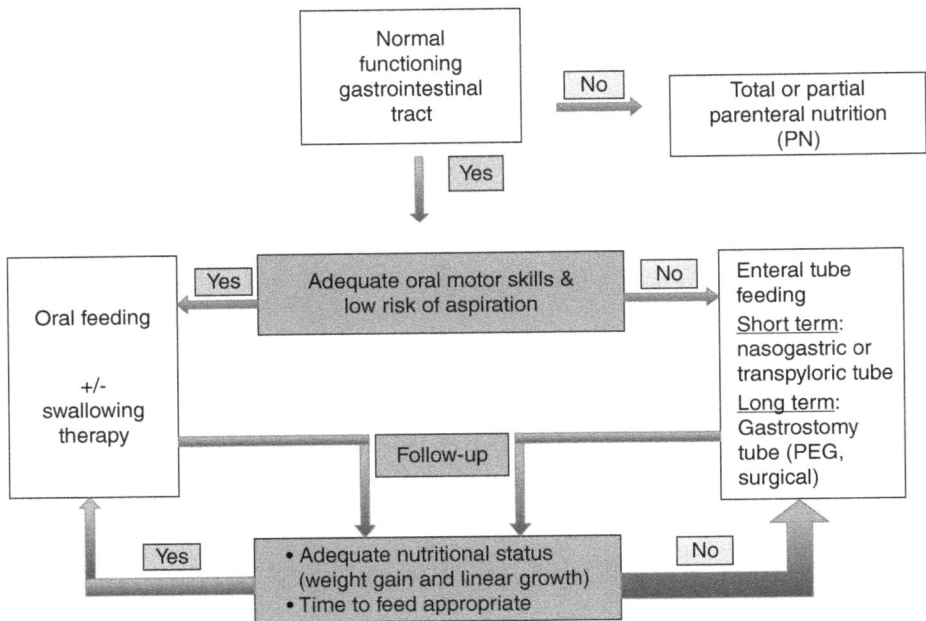

Figure 14.1 Assessment and management of gastrointestinal symptoms in patients with suspected or proven mitochondrial disease

CASE REPORTS

Case Report 14.1

A 16-year-old girl with Leigh syndrome presented to the Emergency Department with severe abdominal pain, diarrhea and vomiting without fever. She was born after an uncomplicated pregnancy as the fifth child of consanguineous parents from an Iraqi family. She reached her developmental milestones within the normal range. At two years of age she presented with a progressive motor disorder with hypotonia complicated by bouts of dystonic posturing. The suspicion of a mitochondrial disease was based on her clinical presentation and elevated lactate levels in blood and CSF. T2 weighted hyperintense lesions in the basal ganglia and cerebellar peduncles on a brain MRI and decreased levels of OXPHOS complex enzymes measured in a muscle biopsy specimen of the quadriceps were observed. Molecular genetic diagnostic testing demonstrated a homozygous mutation in the *NDUFA10* gene that confirmed the diagnosis of Leigh syndrome. During the subsequent years, she became more disabled due to the development of a spastic dystonia and concomitant speech problems. She also suffered from recurrent GI problems with constipation, vomiting, weight loss and swallowing problems requiring several admissions for temporary nutritional therapy with enteral tube feeding.

Physical examination showed a slim girl with a weight of 42 kg and height of 1.65 m (weight-for-height z score of −2.5 SD) with clear neurodevelopmental delay. She could hardly walk due to her dystonia. Her abdomen was minimally distended and by palpation

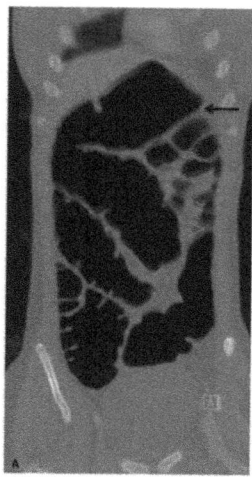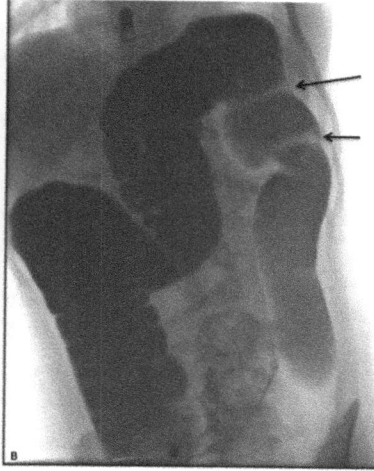

Figure 14.2 A. CT abdomen: Dilated ascending, and transverse colon until the splenic flexura (arrow). The descending colon is difficult to demarcate. Impression of stenotic area in descending colon. Full bladder (bladder retention) and fecal impaction in rectal canal. No signs of volvulus. B. Intestinal obstruction was further analyzed with a barium contrast study for the entire colon. No signs for contrast obstruction. But two caliber changes (arrow) were noted, one directly after the splenic flexure in the descending colon and the other at the beginning of the rectal canal. The combination of MD disease, no signs of mechanical obstruction, diminished peristalsis and haustration of the descending colon together with the dilatation of the proximal colon fits a pseudo-obstruction (Ogilvie syndrome) or CIPO.

diffusely tender. Neurological examination was unchanged compared to previous admissions. An abdominal CT scan was performed and showed a torsion of the sigmoid without signs of ischemia (Figure 14.2). Laboratory investigations showed no signs of inflammation (CRP, WCC), normal electrolytes and a normal arterial blood gas.

Self-Assessment Questions

What is the suspected diagnosis, what additional tests do you need and what is the treatment?

- Viral gastroenteritis
- Appendicitis
- Large bowel obstruction
- Pseudo obstruction (CIPO) syndrome

Answers: Because of the recurrent episodes of abdominal pain with distension of the bowel without any anatomical explanation together with the mitochondrial disease, a pseudo-obstruction syndrome ranks high in the differential diagnosis. Viral gastroenteritis is unlikely due to the recurrent similar previous episodes and her apyrexia. Appendicitis should always be considered in any patient with abdominal pain and remains a largely clinical diagnosis. The abdominal distension and recurrent episodes make this unlikely, however. Large bowel obstruction is possible, but in the absence of previous abdominal surgery, and in previous self-limiting episodes with conservative treatment, pseudo-obstruction should be considered in the first instance.

There is no single lab test to diagnose pseudo-obstruction. In patients with a pseudo-obstruction syndrome, we see severe bowel distension without a mechanical obstruction, and this may require the CT imaging of the abdomen to confirm this. The etiology of CIPO is considered to result from an intestinal myopathy, enteric neuropathy or a combination of the two, but the exact pathophysiology is still unknown in MD patients. CIPO is most often idiopathic, but can also be secondary to an underlying disease, such as a mitochondrial disorder. When there is no evidence of a physical blockage, manometry and full thickness biopsies might help confirm the diagnosis, but the evidence base remains far from clear. In our patient, the CIPO is MD related.

Management: The patient was treated with laxatives and bowel lavage. Because of her persistent swallowing problems due to her dystonia leading to a poor nutritional status, she received a percutaneous gastrostomy for enteral feeding a few months later. Over the next three months, she regained her weight and had only mild GI symptoms left.

Case Report 14.2

A 58-year-old woman was evaluated for weight loss. She had a 38-year history of bilateral ptosis, ophthalmoplegia and pigmentary retinopathy. Her symptoms began aged 20 with a ptosis of her left eye before the development of bilateral ptosis. Three years later, her right eye developed ptosis, followed by the subsequent development of diplopia. She had problems with reading, partly because she had to lift her head, which was tiresome, and partly she was troubled by her diplopia. Based on her clinical phenotype, she was diagnosed with Kearns-Sayre syndrome (KSS) at the age of 28.

At the age of 50, a molecular confirmation of KSS was achieved with the detection of a 4.5kb single deletion of the mtDNA in a muscle biopsy. The deletion was present with a heteroplasmy percentage of 41 percent in muscle and 0 percent (undetectable) in blood. Around this time (aged 50) the patient experienced two episodes of weakness in her left arm, together with feeling that her left arm was clumsy and alienated. Her symptoms fully resolved over a few hours on both occasions. These episodes were classified as stroke-like episodes.

By her next consultation, the patient looked cachectic. She had lost 30 kg in a three-month period and reported swallowing difficulties, nausea and loss of appetite. In parallel, she had been diagnosed as having developed moderate left ventricular cardiac dysfunction and a second-degree aortic valve insufficiency.

Functionally, she is employed part-time in an administrative job where she has noticed that she is becoming easily fatigued. The family history is unremarkable.

Clinical investigations: Physical examination revealed a cachectic (49 kg, 175 cm, BMI: 16.0) woman with dysarthria and a nasal speech. Bilateral ptosis and a complex ophthalmoplegia were present. Mild ataxia in both arms was observed together with pyramidal weakness and spasticity globally. Reflexes were symmetrically brisk (+1), with an extensor plantar response bilaterally. There was a mild (MRC 4/5) proximal decrease in muscle strength. Sensory examination was normal.

Laboratory findings: Blood tests showed a normal hemoglobin level and mean corpuscular volume, lactate – 3.9 mmol/L(ref: 0.5–1.7), pyruvate – 103 mmol/L (ref: 34–103), creatine kinase – 166 IU/L(ref: 0–144), and a complex I deficiency was observed within a muscle

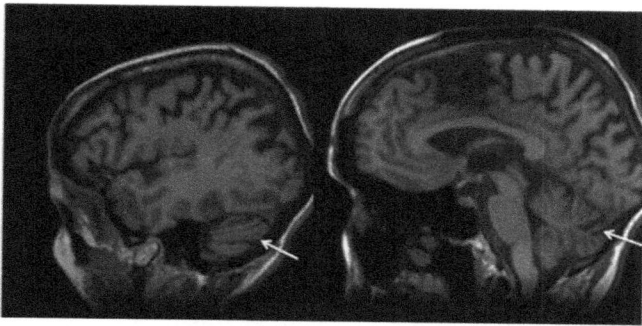

Figure 14.3 Magnetic resonance imaging of the brain. T1 weighted sagittal images. Atrophic cerebellar hemisphere (left arrow). Mild diffuse atrophy of the brain more outspoken in the cerebellar vermis (right arrow) and the pontocerebellar cistern.

biopsy. Histological examination also revealed ragged red fibers in the muscle biopsy, and a brain MRI scan showed infra- and supratentorial dilated CSF-spaces, and diffuse cerebral and cerebellar atrophy. (Figure 14.3)

Self-Assessment Questions

What is the most likely diagnosis?

- Reactive depression
- Malignancy
- A cardiomyopathy
- Neuromuscular dysphagia

Answers: There is no information in the history to suggest a depressive disorder. A malignancy should always be considered in the context of dysphagia, nausea and significant weight loss, and structural lesions should therefore be excluded as part of the investigations. Her left ventricular impairment may contribute to fatigue, though would not explain such dramatic weight loss. The most likely diagnosis here is therefore that the primary pathology is dysphagia with associated additional upper GI dysfunction due to her KSS.

Management: The patient was started on overnight enteral feeding via a percutaneous endoscopic gastrostomy (PEG) tube. In the daytime, the patient was able to continue with a high-caloric, soft food diet. The initial management plan was to optimize the patient's nutritional status and weight. A speech and language therapist supported her specific adjuncts such as drinking water after meals to clear the oropharyngeal area. After six months, the patient's weight increased to 60 kg (BMI: 19.4). Her weight has remained stable subsequently. While her dysphagia persists, and is likely to progress, the immediate management plan has stabilized her nutritional status.

Case Report 14.3

A six-year-old girl presented to the outpatient pediatric clinic with behavioral and eating problems. Her parents had noticed that she did not play with other children and did not seem to talk as much with her teacher. She appeared to have no problems listening or following a discussion, but they felt she had developed problems expressing herself. The patient is the third child of consanguineous parents, although no other family members had similar problems.

A few years later, she was severely cachectic. Extensive metabolic analysis and a CT-scan of the brain revealed no abnormalities. A psychosocial problem was suspected, but a somatic disease could also not be excluded at that time. Psychological and psychiatric treatment seemed to stabilize and even improve some aspects of her behavioral problems, but her cachexia remained.

At the age of 11, she was referred to the pediatric neurology outpatient clinic where her mother told the pediatricians that her learning difficulties hadn't significantly changed over the past few years, but that she was beginning to become more clumsy. She continued to worry about the weight of her daughter and told the pediatricians about profound audible bowel sounds and mild abdominal cramps that her daughter had noticed. She also had developed persistently loose stools, though her appetite and swallow remained normal.

Her disease progressed. By the age of 17, she needed a wheelchair for longer distances due to progressive generalized weakness. Her weight remained extremely low at around 26 kg (BMI: 11.5).

She was admitted to the hospital aged 17 for the commencement of parenteral nutrition.

Clinical investigations: Physical examination showed a skull circumference of 53.5 cm (0 SD), length 148 cm (−0.2 SD) weight 26 kg (weight-for-age −2.5 SD; weight-for-height −3.5 SD) and BMI 12 (−4 SD). Physical examination revealed a slight ptosis, mild myopia and a discrete ophthalmoparesis affecting horizontal and vertical eye movements. She had evidence of significant muscle wasting, with a mild proximal muscle weakness of MRC 4. Her gait was stiff. Her gross motor function measure (GMFM) standing was 97 percent, and walking, running and jumping 92 percent.

Reflexes were globally barely detectable, and plantar responses were flexor. Sensory examination showed a discrete loss of vibration perception below the ankles.

Laboratory investigations included liver function tests (also lipid analysis) and renal function tests, whole blood count, glucose, uric acid, CK and phosphate. A metabolic work-up was completed with plasma amino acids, acylcarnitine profile, glycosylation pattern and very-long-chain fatty acids. A urinary metabolic screen including amino and organic acids, purine and pyrimidine metabolites was done. Cerebrospinal fluid (CSF) analysis of cells, protein and amino and lactic acids levels was also performed.

Abnormal lab results consisted of: CK 213 u/l (n <123 u/l), ASAT 54 u/l (n < 26 u/l), ALAT 111 u/l (n<22 u/l), Creatinine 23 μmol/l (37–80 μmol/l), (GT 81 μmol/l (n <33 μmol/l), LD 545 u/; (n < 434 u/l), lactic acid 3 mmol/l (n <1.7 mmol/l), pyruvate 128 μmol/l (34–103 μmol/l), 3-hydroxybutyrate <0.10 mmol/l(0.06–02 mmol/l). In CSF protein 1.55 g/l(0.18–0.58 g/l), lactic acid 3.6 mmol/l (0.9–2.8 mmol/l), pyruvate 123 μmol/l (60–190 μmol/l). Metabolic tests showed an elevated thymidine 28 μmol/mmol kreat (ref. 0–2) and deoxyuridine levels 45 μmol/mmol kreat (ref. 0–11) in urine.

MRI brain: Supratentorial white matter was abnormal, with the splenial part of the corpus callosum more involved than the genu, and U fibers appear to have been spared. The white matter involvement seems to spread from periventricular regions towards the subcortical regions (Figure 14.4). Corticospinal tracts were also involved and infratentorially, the pons and the mesencephalic regions also showed significant white matter abnormalities. On MRS, the N-acetylaspartate (NAA) peak was low in the white matter area regions, but spared compared to NAA changes seen in demyelinating disease disorders.

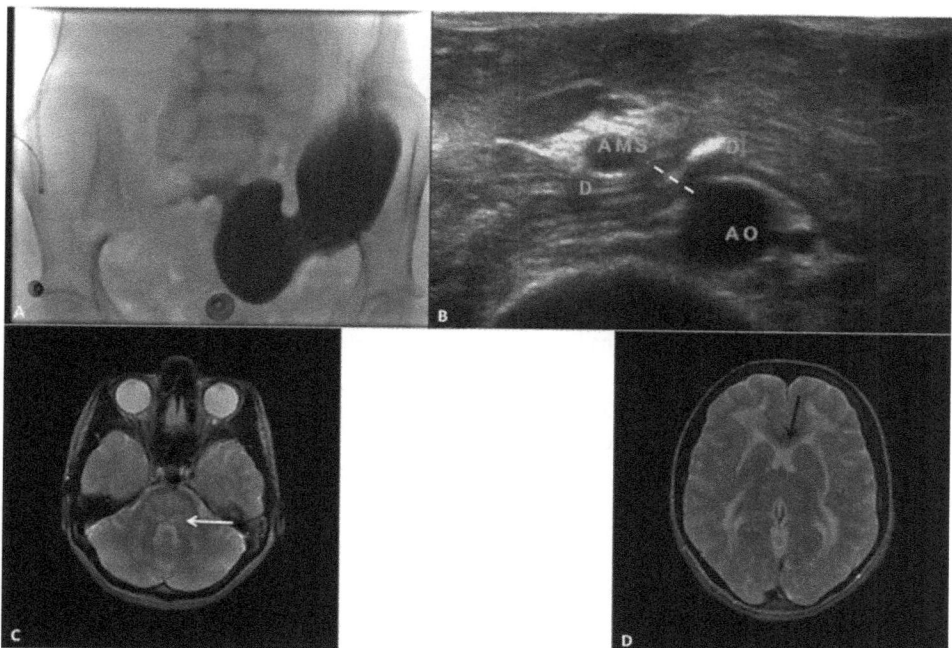

Figure 14.4 Barium contrast study and an abdominal echo of stomach and duodenum demonstrates A. Elongated stomach with the antral part of the stomach in the pelvic area. A delayed passage from stomach toward duodenum. B. Abdominal echo shows a short aortic – mesenterial arterial distance (dashed line). Compatible with SMA syndrome. AMS: mesenteric superior artery, D: duodenum, AO: aortic artery. C. and D.: MRI of the brain with white matter hyperintensities on the T2 weighted axial images. C. Infratentorial changes in the pons and mesencephalic corticospinal tracts (arrow) and D. Supratentorial extensive and diffuse white matter changes with sparing of the U-fibres (white arrow) and part of the callosal body (black arrow).

EEG: Normal background. No signs of epileptiform activity.

Brainstem evoked potentials: Discrete delay at the level of the brainstem. Hearing loss: Right ear: 3kHz: 20 db and left: 3kHz: 10 db.

Electroneurography: Demyelinating sensorimotor polyneuropathy.

Muscle biopsy: Some COX negative fibers. Electron microscopy: no abnormalities.
   Respiratory chain complex activities (percent of normal mean): Complex I = 80 percent, II = 120 percent, III = 106 percent, IV = 182 percent, V = 77 percent, Citrate synthase = 66%.

Genetic analysis: mtDNA analysis from muscle showed no pathogenic mutations. No signs for mtDNA depletion were observed. In blood, no mtDNA deletion was detectable, but in muscle, a low percentage level of multiple mtDNA deletions was observed.

Gastrointestinal studies: Barium studies (X-ray) showed delayed passage from the stomach toward the duodenum. Abdominal ultrasound showed superior mesenteric artery syndrome in which the third and final portion of the duodenum is compressed between the abdominal aorta and the overlying superior mesenteric artery (Figure 14.4B) due to the loss of visceral fat in severe cachexia.

Ophthalmology examination (aged 16): White cristaline depositions. In the periphery of the retina, slight salt-and-pepper-like appearances were also observed.

Cognitive examination (aged 12): This test showed a dysthymic disorder with a learning disorder. Her total IQ was 82 with no difference between nonverbal and verbal abilities.

Self-Assessment Questions

- How would you classify this phenotype?
- What is unusual about the MRI imaging findings in the context of mitochondrial disease?
- Why might the patient have developed superior mesenteric artery syndrome (SMA)?
- What is the most likely diagnosis and how would you confirm this in the patient?

Answers: This patient suffers from a multiple-mtDNA deletion syndrome, characterized by a multisystem phenotype with prominent GI symptoms and a progressive cognitive and motor phenotype in association with prominent and progressive white matter changes. The prominent and progressive subcortical white matter changes that are consistent with demyelination with MRS are unusual in the context of mitochondrial disease (see Chapters 2 and 7). The rapid and profound weight loss on this patient has likely precipitated the superior mesenteric artery syndrome. This condition develops as intraabdominal fat is lost; it causes a reduction in the angle between the SMA and the aorta, which may subsequently compress the third part of the duodenum, causing small bowel obstruction. The elevated thymidine and deoxyuridine levels suggest impairment of thymidine phosphorylase (TP) which is observed in mitochondrial neurogastrointestinal encephalomyopathy syndrome (MNGIE). Sequencing of the *TYMP* gene should be undertaken. In this case, a homozygous pathogenic mutation was detected.

Management

Metabolic tests revealed elevated thymidine 28 μmol/mmol (ref. 0–2) and deoxyuridine levels 45 μmol/mmol kreat (ref. 0–11) in urine pointing to a defect in the enzyme thymidine phosphorylase (TP). The metabolic finding was confirmed by a homozygotic mutation in the *TYMP* gene that encodes thymidine phosphorylase. TP deficiency is associated with mitochondrial neurogastrointestinal encephalomyopathy syndrome (MNGIE), which is the diagnosis in this case. The mtDNA depletion can be seen in muscle and is likely caused by imbalance of nucleotides, in particular the secondary deficiency of dCTP might contribute to the neurodegeneration. The intestinal dysfunction may start in the teenage period and is the cause of death 10 to 20 years later. The current therapies are mainly supportive [8].

Early intervention with allogenic bone marrow transplantation has been effective and prevents further deterioration or supports even partial recovery in patients. The intervention, however, has come with a significant mortality and morbidity, in particular when not applied early in the disease process [9]. This patient eventually received an allogenic bone marrow transplantation (HSCT), but died due to transplant-related complications of graft versus host disease. Liver disease and a history of intestinal pseudo-obstruction might be used as markers of the disease stage and worsen the prognosis for disease course and the chance of a favorable outcome with HSCT.

Erythrocyte-encapsulated thymidine phosphorylase (TP) infusions [10] might contribute to an improvement in the preconditioning of the patient for HSCT and were considered in this case. In the near future, Lentiviral or AAV-mediated gene therapy with the TP gene could be an improvement for allogenic HSCT with lower morbidity and mortality [11].

References

1. Scharfe C, Lu HH, Neuenburg J, et al. Mapping gene associations in human mitochondria using clinical disease phenotypes. *PLoS Comput Biol* 2009 5(4): e1000374.

2. Gabbard SL, Lacy BE. Chronic intestinal pseudo-obstruction. *Nutr Clin Pract* 2013 Jun;28(3):307–316. Review.

3. Amiot A, Joly F, Cazals-Hatem D et al. Prognostic yield of esophageal manometry in chronic intestinal pseudo-obstruction: A retrospective cohort of 116 adult patients. *Neurogastroenterol Motil* 2012 24: 1008–e542.

4. Lee WS, Sokol RJ. Mitochondrial hepatopathies: Advances in genetics, therapeutic approaches, and outcomes. *J Pediatr.* 2013 Oct;163(4):942–948.

5. Christen HJ, Hanefeld F, Kruse E, et al. Foix-Chavany-Marie (anterior operculum) syndrome in childhood: A reappraisal of Worster-Drought syndrome. *Dev Med Child Neurol* 2000 Feb;42(2):122–132. Review.

6. Kornblum C, Broicher R, Walther E, et al. Cricopharyngeal achalasia is a common cause of dysphagia in patients with mtDNA deletions. *Neurology* 2001 May22;56 (10):1409–1412.

7. Chitkara DK, Nurko S, Shoffner JM, et al. Abnormalities in gastrointestinal motility are associated with diseases of oxidative phosphorylation in children. *Am J Gastroenterol* 2003 Apr;98(4):871–877. Review.

8. Yadak R, Sillevis Smitt P, Van Gisbergen MW et al. Mitochondrial neurogastrointestinal encephalomyopathy: From pathogenesis to emerging therapeutic options. *Frontiers in Cellular Neuroscience.* 2017 epub Manuscript ID: 247080.

9. Halter JP, Michael W, Schüpbach M, et al. Allogeneic haematopoietic stem cell transplantation for mitochondrial neurogastrointestinal encephalomyopathy. *Brain* 2015 Oct;138 (Pt. 10):2847–2858.

10. Bax BE, Bain MD, Scarpelli M, et al. Clinical and biochemical improvements in a patient with MNGIE following enzyme replacement. *Neurology* 2013; 81: 1269–1271.

11. Di Meo I, Lamperti C, Tiranti V. Mitochondrial diseases caused by toxic compound accumulation: From etiopathology to therapeutic approaches. *EMBO Mol Med* 2015 Jul 20;7 (10):1257–1266.

Nephrology

Thomas M. F. Connor and Patrick H. Maxwell

Introduction

The kidneys are a pair of organs that lie in the retroperitoneal space and weigh approximately 130 grams each in adult men. The kidneys develop through a series of embryological stages, reaching structural maturity by weeks 32–36 of human gestation. The functional unit of the kidney is the nephron, and each human kidney typically contains around one million nephrons. This number is fixed before birth and the mature kidney is unable to replace lost nephrons. Each nephron consists of a glomerulus, the site of plasma ultrafiltration, and a tubule, where salt and water are reabsorbed.

The kidneys are responsible for three broad aspects of physiological homeostasis. The first is the regulation of body fluid composition; they control the osmolality, salt content and acidity of the extracellular fluid. Second, the kidneys are responsible for the excretion of metabolic waste products, such as urea and creatinine, and foreign substances, such as drugs and toxins. Third, the kidneys have a number of endocrine functions, including the synthesis of renin, erythropoietin and 1,25-dihydroxyvitamin D_3. All these homeostatic functions are impaired by serious renal disease. Normal humans have at least twice the renal capacity necessary for good health, and there are usually no metabolic consequences of progressive renal disease until the glomerular filtration rate (GFR) has fallen below 30 ml/min.

The kidneys are highly metabolically active; despite their small mass, they are responsible for 4–8 percent of the resting body's energy turnover [1]. The kidneys contain a high density of mitochondria, especially in the cortical tubules, where the majority of solute reabsorption occurs [2]. Renal oxygen consumption is directly proportional to the amount of sodium reabsorbed. In view of the dependence of renal function on aerobic metabolism, it is not surprising that impairment of normal mitochondrial function, following insults such as ischemia, drug toxicity and genetic mitochondrial disease, can lead to kidney disease [3].

Kidney disease has not been regarded as a common feature of inherited mitochondrial disease. However, kidney involvement has been underappreciated and detailed characterization suggests renal involvement in up to 50 percent of children with genetic mitochondrial disease [4–6].

Clinical Presentations

Mitochondrial disease can result in four types of chronic kidney problems that may overlap.

Abnormal tubular function: The most common renal abnormality in mitochondrial diseases is impaired reabsorption of components of the ultrafiltrate by the proximal renal tubule, which is referred to as renal Fanconi syndrome when severe. This involves a failure to reabsorb bicarbonate, glucose, phosphate, uric acid, amino acids and low molecular weight proteins. Clinical features include polyuria, polydipsia, dehydration and growth failure in children. Other features include proximal renal tubular acidosis, hypophosphatemic rickets or osteomalacia, with systemic hypokalemia and hyperchloremia. Patients with mitochondrial disease often have a partial (or mild) form of renal Fanconi syndrome. In other individuals, more distal components of the renal tubule may be affected. This can result in a Bartter-like syndrome, with sodium wasting, hypokalemia and abnormal losses of magnesium and calcium into the urine. It can also produce a failure of urinary concentrating ability, with nephrogenic diabetes insipidus.

Malfunction of the capillary filtration barrier in the glomerulus: This results in significant albuminuria. If severe, this presents as nephrotic syndrome, characterized by protein excretion of more than 3.5 g/day in adults or 40 mg/m^2/h in children, in association with edema and hypoalbuminemia. Glomerular leakage is a particular feature of patients with mitochondrial disease due to coenzyme Q_{10} defects and tRNA^{LEU} mutations. On renal biopsy the usual histological lesion associated with proteinuria in mitochondrial diseases is focal segmental glomerulosclerosis (FSGS; see Figure 15.2).

Cystic renal disease: Some genetic mitochondrial disorders are associated with cystic kidneys at birth, related to abnormal renal development.

Reduced overall renal function (often referred to as chronic kidney disease, or CKD): This is usually asymptomatic and identified on the basis of raised plasma creatinine on a routine blood test. Rarely, individuals present with symptoms of end-stage renal disease (ESRD), including reduced urination, fatigue, nausea, anorexia, itching and muscle cramps. In mitochondrial diseases, renal biopsy typically shows tubulointerstitial inflammation, fibrosis and atrophy.

In addition, mounting experimental evidence suggests that mitochondrial dysfunction plays a key role in acute kidney injury (AKI). AKI is common in hospitalized patients, often as part of multi-organ dysfunction in patients with any serious illness. Patients with rhabdomyolysis due to mitochondrial diseases (or other causes) commonly develop AKI as a consequence of myoglobinuria. Many definitions of AKI exist, but all involve a reduction in urine output and elevation of serum creatinine. It is possible that recurrent episodes of subclinical AKI may contribute to chronic kidney disease (CKD).

Clinical Examination

Clinical examination of the renal system is based on an assessment of the normal homeostatic control of salt and water balance. Initial assessment should include measurement of height and weight, inspection for peripheral edema and assessment of the jugular venous pressure. Measurement of blood pressure is a critical part of the renal examination.

Failure of tubular reabsorption can lead to chronic reduction in effective circulating fluid volume (ECFV). This can result in postural hypotension, tachycardia and cool peripheries. Less reliable signs include reduced skin turgor, delayed capillary refill, sunken eyes and dry mucous membranes. More acute reductions in ECFV result in hypotension, tachycardia and reduced urine output, and can contribute to AKI.

More commonly, CKD results in salt and water retention, hypertension and increased extracellular fluid volume. Increased interstitial fluid results in peripheral or generalized edema, with ascites and pleural effusions. Generalized edema is a particular feature of the nephrotic syndrome, and may affect the face and hands if severe or after a period recumbent. Increased central venous pressure can be associated with signs of pulmonary edema, including tachypnea, tachycardia, a third heart sound and bibasal crackles.

Physical findings in CKD vary depending on the severity of kidney failure, and may be within normal limits. Reduced erythropoietin synthesis may lead to anemia, while reduced 1.25-dihydroxyvitamin D_3 can lead to the development of rickets and signs associated with hyperparathyroidism, such as bony tenderness or pseudo-clubbing. Chronic fluid overload and hypertension may lead to left ventricular hypertrophy and a displaced apex beat. Untreated acidosis is associated with increased respiratory rate (Kussmaul breathing), while hypocalcemia may be associated with tetany. Neurological signs in advanced CKD include peripheral neuropathy and changes in mental status, ranging from lethargy to seizures and coma. Severe uremia may be associated with a distinctive fetor, sallow complexion and frosting, metabolic flap and easy bruising.

Examination of the urine is an essential part of routine clinical examination in renal disease and can give vital information within a few minutes. Glycosuria occurs in both diabetes and proximal tubular dysfunction. Elevation of urinary pH (>5.3) in the context of systemic acidosis is indicative of renal tubular acidosis. Urine dipsticks provide a semi-quantitative measurement of proteinuria. Heavy proteinuria is characteristic of nephrotic syndrome, and any proteinuria merits further investigation of renal function. Microscopic hematuria, leukocyte esterase and nitrites can all be detected easily using dipsticks, but are not typically associated with mitochondrial diseases.

Clinical Investigations

Renal imaging is a routine part of assessment of the kidneys; however, there are no characteristic findings in mitochondrial disease. Renal tract ultrasound is the most commonly utilized imaging modality as it is relatively inexpensive and provides a rapid way to assess renal location, contour and size without radiation exposure. Moreover ultrasound can be used safely during pregnancy. The most common ultrasound finding in renal disease is increased cortical echogenicity with loss of the normal cortico-medullary differentiation (see Figure 15.3); however, this carries little prognostic significance on its own.

Other imaging modalities (CT and MRI) may be recommended for further characterization of abnormalities identified on ultrasound. Nuclear scintigraphy can be used to give an accurate measurement of GFR, as well as divided and dynamic excretory function.

Renal biopsy is the single most useful test for the diagnosis of mitochondrial renal disease. It is usually performed as a day-case procedure, percutaneously, using real-time ultrasound guidance. Common indications include proteinuria (protein/creatinine ratio >100 mg/mmol), abnormal or deteriorating renal function and persistent microscopic hematuria. It is essential to include a sample for electron microscopy; in addition to information about the glomerular basement membrane and identification of electron dense deposits, this enables examination of mitochondrial ultrastructure.

Laboratory Investigations

Key laboratory investigations in renal disease are measurements of serum and urine biochemistry. Serum creatinine is the single most useful measurement, and combined with age and sex can be used to estimate the glomerular filtration rate (eGFR). Normal GFR decreases with age, and is lower in women than men. If the creatinine is above the normal range, this always indicates significant impairment of renal function. However, GFR is reduced by more than 50 percent before this occurs, and it is important to appreciate that a creatinine value in the normal range does not imply that the GFR is normal (see Case Report 15.3).

It is important to measure electrolytes in renal disease, including sodium, potassium, bicarbonate, calcium, phosphate, chloride and urate. Albumin should be measured in all patients and will be reduced in patients with nephrotic syndrome. Investigations in CKD should include assessment for possible bone disease (parathyroid hormone, alkaline phosphatase and total vitamin D) and anemia (full blood count and hematinics).

Investigation of the urine includes measurement of proteinuria using the albumin- or protein-creatinine ratio (ACR or PCR) on a spot urine sample. Unlike the urine dipstick, this provides accurate quantification that is not affected by urinary concentration, and it has largely come to replace the 24-hour collection for proteinuria in adult practice.

Investigation for renal Fanconi syndrome should include 24-hour collection of urine for measurement of sodium, potassium, phosphate, urate and amino acids. If a 24-hour collection is not possible, the fractional excretion of these electrolytes can be calculated on a spot sample using the following formula:

$$Fractional\ excretion = \frac{Urine\ electrolyte \times Serum\ creatinine}{Serum\ electrolyte \times Urine\ creatinine}$$

The urinary loss of low molecular weight proteins seen in this condition can be assessed with assays for retinol binding protein (RBP), N-Acetyl-β-D-glucosaminidase (NAG), or neutrophil gelatinase-associated lipocalin (NGAL). Renal tubular acidosis can be confirmed by dynamic measurement of plasma bicarbonate and urinary pH following ingestion of ammonium chloride, or frusemide and fludrocortisone. Likewise, loss of urinary concentrating ability can be assessed by measurement of paired urine and serum osmolalities, or a formal water deprivation test.

Managing Complications

Managing the complications of renal disease is specific to the type of presentation. Close monitoring of all patients with AKI is essential, with optimization of volume status and imaging to exclude obstructive causes of renal impairment. In patients with multisystem mitochondrial cytopathy, renal impairment may well be secondary to multiple etiologies, including reduced fluid intake, catheter-related sepsis, drug toxicity or cardiac dysfunction.

Treatment of renal tubular dysfunction is directed, whenever possible, toward the underlying cause (see Case Report 15.1). Where the cause is hereditary, therapy is directed at the biochemical abnormalities secondary to the renal solute losses and at the bone disease often present in those patients. Patients with acidosis may require large doses of alkali for correction. Where there is associated hypokalemia, the use of potassium citrate, lactate or acetate will correct both hypokalemia and acidosis. Some patients may require sodium supplementation, either as sodium bicarbonate or sodium chloride. Magnesium supplementation may be required. Correction of hypokalemia may reduce the polyuria, since potassium depletion can itself impair the concentrating ability of the distal tubule.

Hypophosphatemia is multifactorial, but can be treated with oral phosphate with the goal of normalizing serum phosphate levels. Many patients with Fanconi syndrome will require supplemental vitamin D (usually 1.25-dihydroxycholecalciferol) for adequate treatment of rickets and osteomalacia. Supplemental calcium may be indicated for hypocalcemia after supplemental vitamin D is started. The proteinuria and hyperuricosuria in this condition do not require treatment, but carnitine supplementation may improve muscle function and lipid profiles.

Treatment of CKD is directed toward slowing progression, with tight control of blood pressure and reduction of proteinuria using blockade of the renin-angiotensin system, and correction of metabolic complications. There are evidence-based guidelines on the management of CKD [7]. Acidosis should be treated with oral sodium bicarbonate, renal bone disease can be corrected using 1-hydroxycholecalciferol and phosphate binders for hyperphosphatemia, and anemia corrected using iron supplementation and erythropoietin. Ultimately, progressive CKD requires timely and informed consideration of renal replacement therapy, including dialysis and transplantation.

CASE REPORTS

Case Report 15.1

A 47-year-old Caucasian male was referred for investigation of proteinuria and a rise in serum creatinine. He was generally well, but gave a history of generalized aches and pains, particularly affecting the ribs and both heels, for the previous few months. He had been diagnosed with HIV in 1985 and had been on multiple treatment regimens of highly active antiretroviral therapy due to viral resistance. Disease control was now excellent with undetectable viral load and normal CD4 count. He had no other medical problems, and specifically did not suffer from diabetes mellitus or hypertension. His medications included tenofovir, didanosine, ritonavir, lopinavir, atorvastatin and occasional ibuprofen. He smoked tobacco and occasional cannabis, but did not drink alcohol. Childhood development was normal and there was no family history of renal disease.

Physical examination was unremarkable and his blood pressure was 124/65 mm Hg. Urinalysis by dipstick showed pH 5.6, 1+ protein, 1+ blood, and 2+ glucose.

Investigations (Table 15.1):

Analyte	Result	Normal range
Sodium	139 mmol/L	(133–146)
Potassium	3.4 mmol/L	(3.5–5.3)
Creatinine	134 umol/L	(60–125)
Estimated GFR	55 ml/min/1.73 m^2	(>60)
Chloride	110 mmol/L	(95–108)
Bicarbonate	20 mmol/L	(22–29)
Calcium	2.23 mmol/L	(2.15–2.60)
Phosphate	0.51 mmol/L	(0.8–1.5)
Magnesium	0.67 mmol/L	(0.70–1.00)

(cont.)

Analyte	Result	Normal range
Alkaline phosphate	176 IU/L	(30–130)
Albumin	40 g/L	(35–50)
Urate	260 umol/L	(110–420)
Glucose	7.2 mmol/L	(3.0–7.8)
Parathyroid hormone	4.7 pmol/L	(1.1–6.8)
Vitamin D	34.7	(70–150)
Urine PCR	186 mg/mmol	(<20)

Clinical Investigations

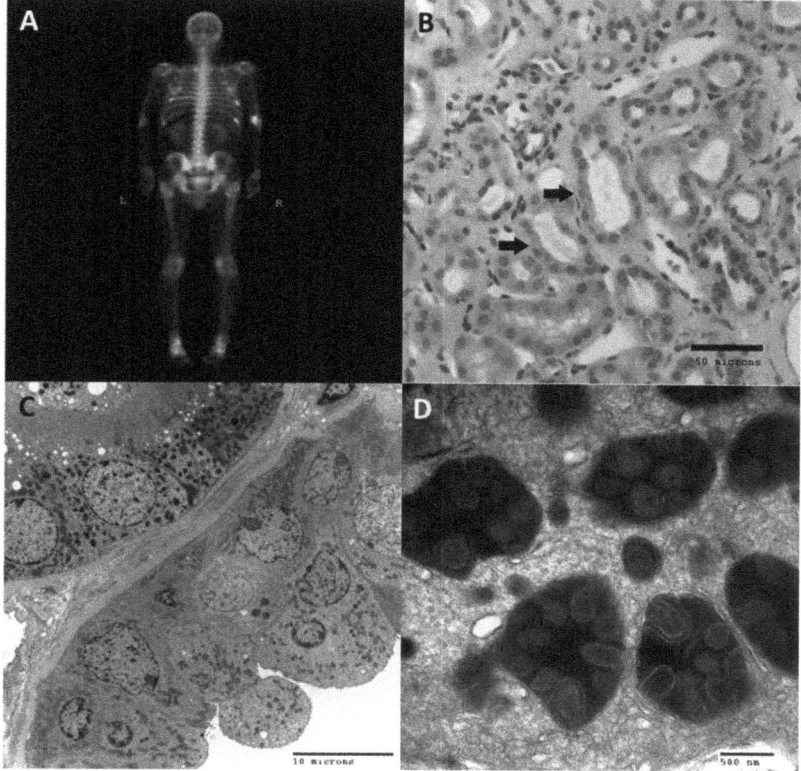

Figure 15.1 (**A**) bone scan showing evidence of osteomalacia in adult patient with tenofovir-induced Fanconi syndrome; (**B**) light micrograph of renal biopsy showing acute tubular injury with epithelial cell flattening and dedifferentiation (arrowheads); (**C** and **D**) electron micrograph of renal biopsy showing abnormal mitochondria, with a highly electron-dense, condensed matrix, and occasional loss or distortion of cristae, in the epithelial cells of proximal tubules. (A black and white version of this figure will appear in some formats. For the color version, please refer to the plate section.)

Self-Assessment Questions

- What is the most likely diagnosis, and disturbance of which electrolytes is diagnostic?
- Which part of the kidney is affected?
- Why is the exclusion of diabetes mellitus important?
- What other conditions can give rise to this presentation?

Answers

- The diagnosis of Fanconi syndrome was confirmed on a 24-hour urine collection which demonstrated an increased fractional excretion of phosphate and urate, and generalised aminoaciduria.
- Increasing evidence suggests that TDF is directly toxic to mitochondria in the proximal tubule [8, 11]. The exact mechanism of pathogenesis is unknown; however studies have shown evidence of mitochondrial ultrastructural abnormalities in parallel with depletion of mitochondrial DNA.
- In order to diagnose glycosuria due to tubular dysfunction, it is important to exclude diabetes mellitus, which causes glycosuria by physiological saturation of tubular transporters.
- Mitochondrial toxicity, including myopathy, lactic acidosis, and hepatic steatosis, has been described with other nucleoside reverse transcriptase inhibitors. Moreover, Fanconi syndrome also occurs in patients exposed to other mitochondrial toxins, including aminoglycosides, valproate, and ifosphamide [3].

Approach

Fanconi syndrome is the most common presentation of patients who have pathogenic mutations in mitochondrial DNA [12]. One third of cases present in the neonatal period, and most before the age of two years. Most patients exhibit a partial Fanconi syndrome, with isolated aminoaciduria, renal tubular acidosis, glycosuria, or hypercalciuria [4]. All patients reported so far had various extra-renal symptoms (hearing loss, neurological symptoms, growth retardation, or diabetes mellitus).

Drug-induced renal Fanconi syndrome is not uncommon and has been reported with several antiretroviral medications, including tenofovir (TDF) [8]. Risk factors for the development of TDF-induced toxicity include older age, low body weight, pre-existing decrease in kidney function, and concomitant use of nephrotoxic medications such as ritonavir or NSAIDs. Whilst severe TDF-associated Fanconi syndrome is rare, affecting <0.1% of patients, evidence of milder tubular dysfunction has been found in up to 22% [9]. Reduced GFR has also been noted in these patients due to inhibition of normal tubular secretion of creatinine [10].

This patient was taken off tenofivir and started on a combination of lamivudine, raltegravir, darunavir, and ritonavir. He was also started on vitamin D replacement with intramuscular ergocalciferol.

Case Report 15.2

A nine-year-old girl was referred with proteinuria and edema. Proteinuria had been identified at the age of five, but despite prolonged courses of high-dose steroids and ciclosporin, it remained in excess of 3 g/day. She had a normal birth and normal developmental milestones. Her medications included furosemide and enalapril. She had two older brothers and one younger sister who were well. Her parents were first cousins and were both well.

Physical examination showed height 122 cm (third centile), weight 28 kg (50th centile) and blood pressure was 104/54 mm Hg. There was pitting edema to the sacrum. Cardiac examination revealed a displaced, heaving apex, with an S3 gallop and ejection systolic murmur. Fundoscopic examination revealed mild optic atrophy. Urinalysis by dipstick showed pH 6.0, 4+ protein, trace blood and negative glucose.

Investigations (Table 15.2):

Analyte	Result	Normal range
Sodium	136 mmol/L	(133–146)
Potassium	3.8 mmol/L	(3.5–5.3)
Creatinine	147 umol/L	(60–125)
Estimated GFR	30 ml/min/1.73 m^2	(>60)
Albumin	13.5 g/L	(35–50)
Urine PCR	340 mg/mmol	(<20)

Given these findings, the patient underwent a renal biopsy. The patient developed progressive renal failure and required hemodialysis one year later. Fifteen months later, the patient developed seizures with focal EEG abnormalities, especially in the left occipital region. Blood creatinine kinase and lactate levels remained normal; cerebrospinal fluid lactate, serum amino acids and urinary organic acids were also normal. MRI demonstrated diffuse cerebral atrophy, mild cerebellar atrophy and bilateral lesions in the cingulate cortex and subcortical area. A muscle biopsy was performed.

Clinical Investigations

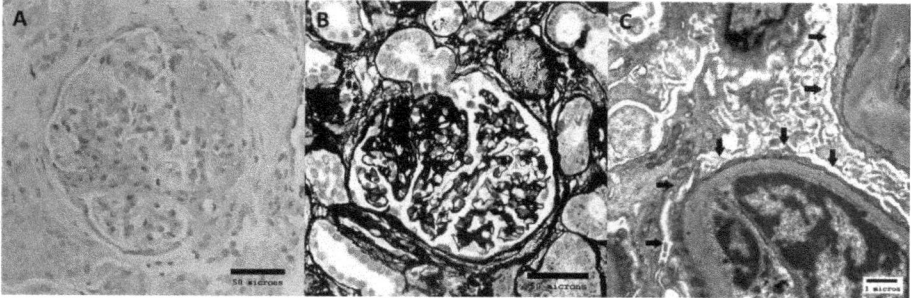

Figure 15.2 (**A** and **B**) Renal biopsy showing segmental sclerosis of a glomerulus by light microscopy (hematoxylin and eosin stain, **A**, and silver stain, **B**); this lesion was found to affect 5 of 40 glomeruli samples, consistent with a diagnosis of focal segmental glomerulosclerosis (FSGS); (**C**) electron microscopy of glomerular epithelial cells (podocytes) showing extensive foot process effacement (arrowheads), which results in malfunction of the capillary filtration barrier in the glomerulus and the clinical finding of proteinuria. (A black and white version of this figure will appear in some formats. For the color version, please refer to the plate section.)

Self-Assessment Questions

- What is the diagnosis?
- Which part of the kidney is affected?
- What is the likely inheritance?
- What treatments are available for this condition?

Answers

- The clinical diagnosis of steroid-resistant nephrotic syndrome (SRNS) due to focal segmental glomerulosclerosis (FSGS) was confirmed on renal biopsy. The underlying cause of FSGS was shown to be an inherited mitochondrial disease by measurement of coenzyme Q_{10} (CoQ_{10}) concentration.
- Damage to visceral epithelial cells within the glomerulus, called podocytes, results in FSGS.
- The underlying cause of disruption of CoQ_{10} biosynthesis was identified by whole-exome sequencing, which demonstrated compound heterozygous mutations in *ADCK4* [13].
- Patients with primary CoQ_{10} deficiency respond to oral CoQ_{10} supplementation (30 mg/kg/day), in contrast to most hereditary mitochondrial disorders, for which there is no effective treatment [13–15].

Clinical Approach

The level of CoQ_{10} was 13 mg/g in fresh muscle tissue (control mean ± SD, 32 ± 12) and 24 ng/mg protein in fibroblasts (control mean ± SD, 105 ± 14). The ratio of quinone-dependent enzymatic activities, which are used to detect unbalanced respiratory chain enzyme functions, (complex I and III, complex II and III, and complex III) were significantly reduced.

Treatment begun within the first 12 months of life has been shown to resolve oedema and significantly reduce proteinuria [16]. Early diagnosis of CoQ_{10} deficiency is critically important because it may not be possible to normalise renal function once CKD has developed. However CoQ_{10} supplementation may well improve extra-renal features including encephalomyopathy.

Case Report 15.3

A 24-year-old male was referred for investigation of renal impairment, which had developed after treatment with acitretin. His medical history included psoriasis and atopic dermatitis, hypertension and bullous lung disease with recurrent pneumothoraxes. He was the middle of six children born to English parents in a non-consanguineous relationship. He had a normal birth and normal milestones, and had always been good at sport. His regular medications included lisinopril and bendroflumethiazide. He drank alcohol in binges, with 15–20 units at a single session, and smoked tobacco and cannabis only.

His two older sisters were both deaf, but a younger brother and younger sister were not affected. One sister also suffered with recurrent fetal loss, gestational diabetes, preeclampsia and persistent proteinuria. A second brother had died at the age of three weeks. The two older sisters had three children between them, who were well although one was reported to have possible attention deficit hyperactivity disorder.

The patient's mother has sensorineural deafness, diabetes, hypertrophic cardio-myopathy and macular dystrophy. She presented with small kidneys on ultrasound and started dialysis aged 53. One of her three sisters has diabetes, deafness and a stroke, and her brother possibly also suffers with deafness. The patient's father is 58 years old and well. The father's three sisters all had emphysema, but his six brothers are well.

Physical examination revealed a normal neuromuscular phenotype, with height and weight on the 10th centile for his age. His blood pressure was 142/84 mm Hg. A detailed neuromuscular examination including fundoscopy was normal, although the pupils were not dilated. Urinalysis by dipstick showed pH 5.2, 1+ protein, negative blood and negative glucose.

Investigations (Table 15.3):

Analyte	Result	Normal range
Sodium	137 mmol/L	(133–146)
Potassium	4.4 mmol/L	(3.5–5.3)
Creatinine	115 umol/L	(60–125)
Estimated GFR	77 ml/min/1.73 m^2	(>60)
Albumin	43 g/L	(35–50)
Chloride	106 mmol/L	(95–108)
Bicarbonate	26 mmol/L	(22–29)
Calcium	2.42 mmol/L	(2.15–2.60)
Phosphate	0.89 mmol/L	(0.8–1.5)
Alkaline phosphate	50 IU/L	(30–130)
Creatinine kinase	134 IU/L	(40–320)
Albumin	40 g/L	(35–50)
Urine PCR	14 mg/mmol	(<20)

Clinical Investigations

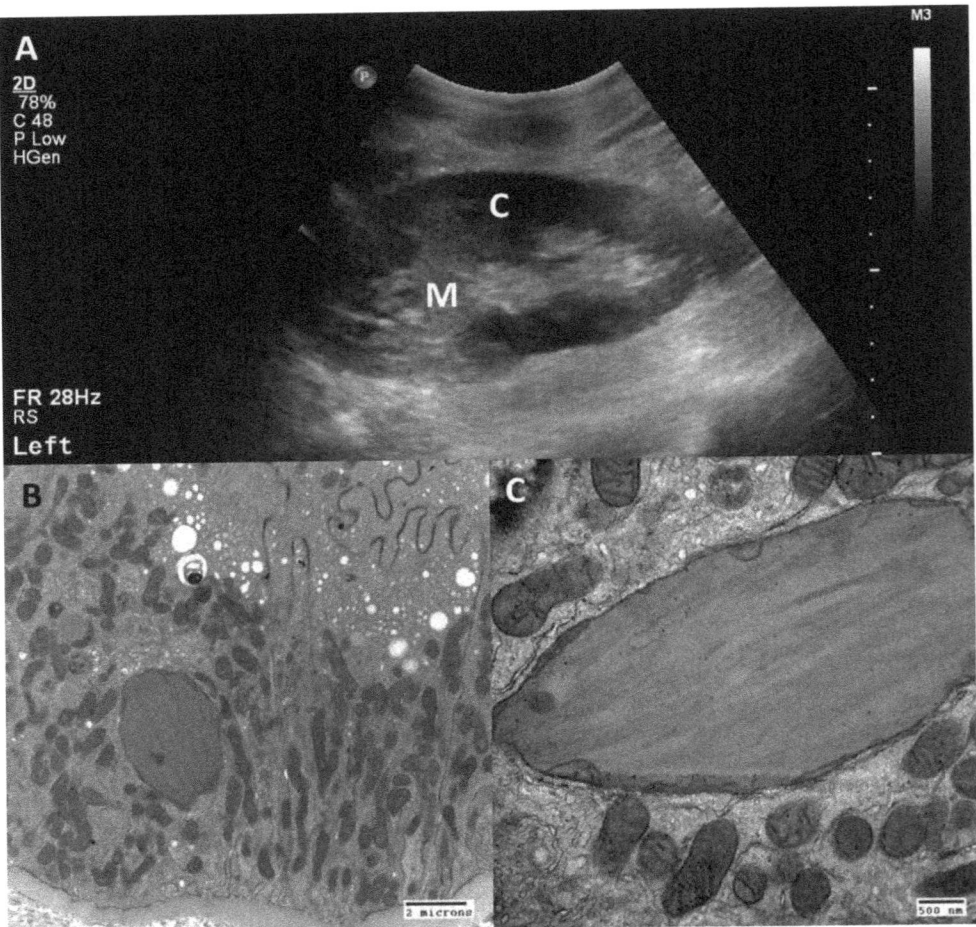

Figure 15.3 (**A**) Renal ultrasound scan showing preservation of normal differentiation between cortex (C) and medulla (M) despite mildly abnormal renal function; (**B** and **C**) electron micrograph of renal biopsy showing occasional giant mitochondria in tubular epithelial cells.

Self-Assessment Questions

- What is the likely diagnosis?
- Which part of the kidney is affected?
- What is the likely inheritance?
- What are the other extra-renal manifestations of this condition?
- What treatments are available for the renal manifestations of this condition?

Answers

- The diagnosis is a mitochondrial tubulopathy causing CKD. FSGS is the most common renal lesion seen in adult patients carrying the m.3243A>G mutation [17]; however a number of families predominantly express other features including tubulointerstitial nephritis and bilateral enlarged cystic kidneys [18]. The clinical course in this condition is heterogeneous, with progression to ESRD between the second and sixth decade.
- Renal biopsy showed abnormal mitochondria in tubular cells.
- Sequencing of mitochondrial DNA revealed the m.3243A>G mutation.
- Extra-renal manifestations include early-onset deafness or diabetes.
- Treatment of renal impairment should follow guidelines for CKD [7], as discussed above. Renal transplantation remains the best treatment for ESRD despite the increased risk of post-transplant diabetes.

Clinical Approach

The patient's renal function was monitored in clinic for 6 months after stopping acitretin therapy with no improvement. It is worth noting that although his creatinine and estimated GFR are within the normal range, a creatinine of 115 represents significant renal impairment for his age and sex. He therefore underwent a renal biopsy. Despite the abnormal mitochondria in tubular cells, the patient did not have features of the Fanconi syndrome. Following the renal biopsy he was referred to audiology and neurology. He underwent a muscle biopsy which was not diagnostic of a myopathy, but which did show cap-like sub-sarcolemmal accumulation of mitochondria and unusually thick small arteries.

The family history is one of the most salient features in this case. As is often the case, this was under-appreciated prior to the patient's renal biopsy. There are other hereditary renal conditions associated with early-onset deafness or diabetes, such as Alport syndrome and autosomal dominant tubulointerstitial kidney disease (ADTKD), which may be more familiar to nephrologists [19, 20]. Extra-renal manifestations of the m.3243A>G mutation may not be evident at the time of presentation with kidney disease, and this case illustrates the importance of taking a complete family history.

References

1. Bullock, J., J. Boyle and M. B. Wang. *NMS Physiology* 578. 2001: Lippincott Williams & Wilkins. 853.

2. Rahman, S. and A. M. Hall. Mitochondrial disease – an important cause of end-stage renal failure. *Pediatr Nephrol* 2013. **28** (3):357–361.

3. Hall, A. M. and R. J. Unwin. The not so "mighty chondrion": emergence of renal diseases due to mitochondrial dysfunction. *Nephron Physiol* 2007. **105**(1): 1–10.

4. Martin-Hernandez, E., et al., Renal pathology in children with mitochondrial diseases. *Pediatr Nephrol* 2005. **20**(9): 1299–1305.

5. Che, R., et al., Mitochondrial dysfunction in the pathophysiology of renal diseases. *Am J Physiol Renal Physiol* 2014. **306**(4): F367–78.

6. Emma, F., et al., Renal mitochondrial cytopathies. *Int J Nephrol* 2011. **2011**: 609213.

7. NICE Chronic kidney disease: Early identification and management of chronic kidney disease in adults in primary and secondary care. *Clinical Guideline* **182**, 2014. 65.

8. Hall, A. M., et al., Tenofovir-associated kidney toxicity in HIV-infected patients:

A review of the evidence. *Am J Kidney Dis* 2011. **57**(5): 773–780.

9. Labarga, P., et al., Kidney tubular abnormalities in the absence of impaired glomerular function in HIV patients treated with tenofovir. *AIDS*, 2009. **23**(6): 689–696.

10. Cooper, R. D., et al., Systematic review and meta-analysis: Renal safety of tenofovir disoproxil fumarate in HIV-infected patients. *Clin Infect Dis* 2010. **51**(5): 496–505.

11. Hall, A. M., Update on tenofovir toxicity in the kidney. *Pediatr Nephrol* 2013. **28**(7): 1011–1123.

12. Niaudet, P. and A. Rotig. The kidney in mitochondrial cytopathies. *Kidney Int* 1997. **51**(4): 1000–1007.

13. Ashraf, S., et al. ADCK4 mutations promote steroid-resistant nephrotic syndrome through CoQ_{10} biosynthesis disruption. *J Clin Invest* 2013. **123**(12): 5179–5189.

14. Montini, G., C. Malaventura and L. Salviati, Early coenzyme Q_{10} supplementation in primary coenzyme Q_{10} deficiency. *N Engl J Med* 2008. **358**(26): 2849–2850.

15. Rotig, A., et al., Quinone-responsive multiple respiratory-chain dysfunction due to widespread coenzyme Q_{10} deficiency. *Lancet* 2000. **356**(9227): 391–395.

16. Salviati, L., et al., Infantile encephalomyopathy and nephropathy with CoQ_{10} deficiency: A CoQ_{10}-responsive condition. *Neurology* 2005. **65**(4): 606–608.

17. Doleris, L. M., et al., Focal segmental glomerulosclerosis associated with mitochondrial cytopathy. *Kidney Int* 2000. **58**(5): 1851–1858.

18. Guery, B., et al., The spectrum of systemic involvement in adults presenting with renal lesion and mitochondrial tRNA(Leu) gene mutation. *J Am Soc Nephrol* 2003. **14**(8): 2099–2108.

19. Eckardt, K. U., et al., Autosomal dominant tubulointerstitial kidney disease: Diagnosis, classification, and management – A KDIGO consensus report. *Kidney Int* 2015.

20. Jansen, J. J., et al., Mutation in mitochondrial tRNA(Leu(UUR)) gene associated with progressive kidney disease. *J Am Soc Nephrol* 1997. **8**(7): 1118–1124.

Psychiatry

Laurence A. Bindoff

Introduction

Mitochondrial diseases are common and highly variable. As explained in the chapters of this book, mitochondria are the primary producers of energy for each and every cell, excluding red blood cells (RBC), and as such, vital for their respective functions. Diseases due to mitochondrial dysfunction frequently involve the central nervous system (CNS) and features of cerebral involvement can include slowly progressive disturbances of brain function, e.g. encephalopathy, cognitive decline, mental retardation, psychiatric illness; episodic disturbance such as epilepsy; or acute functional disturbance such as stroke-like episodes or indeed psychiatric disturbance. The frequent involvement of the CNS has been interpreted as reflecting the high dependence of neurons on readily available energy; another possibility, however, is that symptoms originating from the CNS are among the most alarming for the patient, and those that will most quickly initiate contact with the medical profession.

In this chapter, I will examine the known psychiatric manifestations of mitochondrial disease. Before starting, I should clarify that in the context of this work, mitochondrial disease means that arising from abnormalities of the respiratory chain and its assembly/maintenance/function, or defects affecting the mitochondrial DNA (mtDNA) or the proteins involved in its homeostasis. I will not cover fatty acid oxidation defects, or other metabolic pathways that have their location wholly or partly within mitochondria.

Based on current literature and personal experience, psychiatric manifestations of mitochondrial dysfunction occur predominantly in patients with multisystem disease, particularly those with other forms of major CNS involvement. There are few reports of pure, isolated psychiatric disease and even in those that are available, the vast majority of patients demonstrate the presence of a long-standing, albeit unrecognized systemic disease. It should also be remembered that deafness and visual impairment are common in patients with mitochondrial disease affecting the nervous system, and therefore that sensory deprivation itself can be a potent precipitant of a psychiatric disturbance. Further, any chronic and progressive disease carries with it the pressure for psychological adjustment that in predisposed individuals may manifest psychiatrically. Last, many of the drugs used to treat psychiatric disease can also affect mitochondrial function, and this must be considered both in those with unknown etiology and in those with known mitochondrial disease that presents psychiatrically.

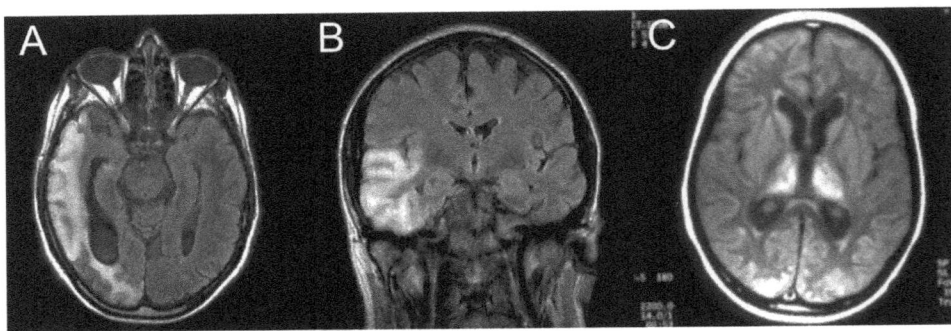

Figure 16.1 MRI scans from a) patient with MELAS and m.3243A>G; b) patient with POLG and p. W748S.
A and B: axial and coronal T2-FLAIR images respectively from a patient with MELAS (m.3243A>G) showing a large, stroke-like lesion in the right temporal lobe. C: axial proton-density weighted scan of a patient with POLG disease (p. W748S homozygous) showing bilateral thalamic and cortical occipital lesions.
Lesions in the lateral temporal lobe are common in MELAS. They often spread from here to the occipital lobes and can involve the auditory association cortex. In *POLG*-related disease, the occipital lobes are areas of predilection and a common site for what are called stroke-like episodes. Patients with lesions here have presented with formed and unformed visual hallucinations that most likely have an epileptic origin.

Psychiatric Disease and Mitochondrial Dysfunction

The prevalence of psychiatric illness in patients with mitochondrial disease is unknown and only a handful of studies compares the incidence of psychiatric disorders in patients with these diseases with the general population, or with other chronic disease. Reviewing the current literature identifies a small number of reports describing patients with a previously unknown mitochondrial disease and a primary psychiatric presentation [1]. In other studies, the authors have attempted to identify mitochondrial dysfunction in patients with pure psychiatric disease by measuring mitochondrial respiratory chain activities in a more available tissue such as skeletal muscle [2]. The remaining and largest number of publications are hypothesis driven, based on the thesis that mitochondrial dysfunction, caused by one of a variety of insults (e.g. free radicals, mtDNA polymorphisms, mtDNA haplotype etc.) might be involved in the causation of major psychiatric diseases such as schizophrenia [3]. Firm evidence for a primary mitochondrial role in these diseases is, however, lacking. Last, we are confronted by the usual problem of definition: for example, a patient with major cerebral involvement and stroke-like episodes due to MELAS who develops cognitive decline and coexistent depression. This is not a purely psychiatric disorder, and usually will not present first to a psychiatrist. Nevertheless, the psychiatric element is important to recognize and treat.

Given the major involvement of the brain in mitochondrial disorders, it is perhaps not surprising that these patients manifest psychiatric features. It must be remembered, however, that the cerebral involvement itself (Figure 16.1) can generate features that may be interpreted as psychiatric. For example, visual phenomena such as flickering lights and colors, formed hallucinations and visual loss occur with involvement of the visual association cortex, and auditory hallucinations can occur in temporal lesions. All are recognized epileptic phenomena particularly with epilepsia partialis continua or subclinical status.

Types of psychiatric presentation (Table 16.1): Mitochondrial dysfunction can present with the whole range of psychiatric symptoms from anxiety to depression and less commonly

Table 16.1 Common psychiatric manifestations of mitochondrial disease

1. Depression
2. Anxiety
3. Bipolar disease
4. Other psychoses

Of these, depression appears the most commonly reported.

psychosis [4, 5]. In an excellent review of adult patients who presented first with their psychiatric disease, Anglin and colleagues described 12 individuals who manifested mostly depression and anxiety as their primary psychiatric disturbance. They also reviewed the previously reported cases with psychiatric diagnoses [6] and identified that many of these too had psychiatric symptoms, often predating the diagnosis of mitochondrial disease. While all of the patients they described had elements in their clinical histories and examinations suggesting a chronic and ongoing nonpsychiatric disorder, the diagnosis of mitochondrial disease was delayed in these individuals, often by several years. If one excludes fatigue, depression, with or without psychotic features, appears to be the major psychiatric manifestation of mitochondrial disease, at least in this group of adult patients [6]. Mitochondrial disease could, however, be associated with all other types of psychiatric illness including anxiety, major psychosis and bipolar disorder [1].

Mitochondrial defects and psychiatric disease: Mitochondrial disease can be caused by both mtDNA mutations and nuclear gene defects, and patients with both are reported with psychiatric disease, again most often a major depressive illness [6]. Interestingly, the two main mitochondrial diseases that comprise the majority of cases described in this article are MELAS (usually, but not exclusively caused by the m.3243A>G) and patients with polymerase gamma (*POLG*) mutations. These disorders share several similarities: both are common and both are associated with acute or chronic changes in the brain. The chronic element manifests as a progressive encephalopathy complicated by acute episodes of what are termed "stroke-like" episodes. In the syndrome of MELAS, these predominantly affect the temporal lobe with extension into the occipital lobes, while in *POLG* disorders, these are more occipital and subsequently spread to the temporal lobe [7, 8]. In both, involvement of the cerebral cortex gives an aggressive epilepsy, which in the case of *POLG* disorders leads to death (in status epilepticus in >50% percent of cases), but in MELAS often burns out after an intense period of activity. In addition, MELAS patients will often have associated sensory deprivation due to sensorineural deafness and both *POLG* disorders and MELAS are associated with visual impairment in which visual hallucinations occur, and a progressive and global cognitive disturbance that in itself can predispose to psychiatric disease.

A similar clinical picture can be seen in other, less common mtDNA and nuclear gene defects, including the cortical involvement and the predisposition to epilepsy, as well as potential for stroke-like phenomena. Neuronal damage clearly exposes these patients to the risk of psychiatric illness, and at present, there is no evidence to suggest that patients with any form of mitochondrial disease develop pure psychiatric disease without such involvement.

Two other potentially psychiatric features of mitochondrial disease warrant special consideration – fatigue and anorexia. Fatigue is common in mitochondrial disease and can reflect the muscle involvement that limits a patient's exercise capacity, or the CNS involvement that blunts their desire to participate in activities. Despite multiple attempts in different laboratories, however, mitochondrial involvement in the disorder known as chronic fatigue syndrome has never been convincingly demonstrated. Anorexia is also common in any form of chronic disease, but in some mitochondrial diseases it may also be very prominent, and this is best exemplified by the syndrome of MELAS, particularly that caused by the m.3243A>G mutation. Patients with a significant disease load early in life are often small and thin and appear unable to gain weight. In addition, there can also be problems with recurrent vomiting that limit the patients' ability to maintain nutrition. These patients will often also have hearing difficulties and myalgia.

The syndrome of recurrent vomiting can have a psychiatric basis, but it has also been described with mitochondrial disease [9]. In the reported cases, this appears on a background of systemic involvement and investigation should reveal the cause [10]. It is worth noticing also that gastrointestinal symptoms are common in patients with mitochondrial disease and often overlooked [11].

Cognitive decline is a major feature of long-standing cerebral involvement caused by mitochondrial disease. In both MELAS and *POLG*-related disease, patients who survive the initial onslaught of the disease often develop cognitive loss. In patients with *POLG* mutations, this is often significant [12] and often associated with mood changes, including periods of apathy and rage and depression.

The risk of psychiatric illness in those with mitochondrial disease: In a study that looked at the occurrence of psychiatric illness in patients with proven mitochondrial disease, Fattal and colleagues found a high occurrence of psychiatric comorbidity with a lifetime prevalence of 54 percent for major depressive illness, 17 percent bipolar disorder and 11 percent panic disease [5]. This study analyzed patients with known mitochondrial disease albeit including several having other defects including carnitine deficiency, cobalamin deficiency and fatty acid oxidation defects. In a similar study, the lifetime prevalence of psychiatric disease was estimated to be 47 percent in a group of patients with proven mtDNA disease [1]. These authors also compared mitochondrial disease patients with another disease group (with Hereditary Sensorimotor Neuropathy [HSMN]) and showed that those with mitochondrial disease had a higher risk of developing psychiatric disease. While the two disorders are not strictly comparable and HSMN has no cerebral involvement, the study confirms the risk of psychiatric illness faced by patients with mitochondrial diseases.

Investigating patients with suspected mitochondrial psychiatric illness: Perhaps the major obstacle responsible for retarding diagnosis of mitochondrial psychiatric illness is the lack of awareness of this diagnostic category among psychiatrists and other physicians to whom these patients are first referred. Mitochondrial disease expression is also highly varied and difficult enough even for those with long experience to recognize. Table 16.2 provides some suggestions to increase awareness of a possible mitochondrial disorder that I have adapted from several other publications.

Once the possibility of a mitochondrial etiology has been raised, investigations should be performed to define the precise nature of the disorder. In many cases, the simplest solution

Table 16.2 Clues to the presence of mitochondrial psychiatric disease

For those physicians caring for psychiatric patients, these guidelines are meant as a prompt, something to keep in mind in order to ask the question "can this patient have mitochondrial disease?" Based on current practice, one of the most important investigations is magnetic resonance imaging (MRI) of the brain. This will in most cases provide a major clue to the presence of a mitochondrial disorder.

Adapted from [6] and own experience.

1. Personal history
   a. Mitochondrial psychiatric disease occurs on a background of already existing systemic disease.
   b. The presence of the following features should make one stop and think:
      i. Deafness
      ii. Diabetes mellitus
      iii. Epilepsy, especially an episode of status epilepticus
      iv. Coexistent myopathy, particularly the presence of external ophthalmoplegia
      v. Combinations of these are even stronger indicators of potential mitochondrial etiology

2. Family history
   a. A first-degree relative (esp. mother) with:
      i. Deafness
      ii. Diabetes mellitus
      iii. Epilepsy, especially an episode of status epilepticus
      iv. Coexistent myopathy, particularly the presence of external ophthalmoplegia
      v. Combinations of such

3. Investigation
   a. MRI showing white matter involvement with predilection for occipital and/or temporal lobes

4. Treatment
   a. Failure to respond to two different adequate trials of treatment
   b. Worsening during or improvement following withdrawal of psychotropic drug

will be to refer the patient to a center that specializes in the study of mitochondrial function/genetics. Measuring biomarkers such as lactate and other organic acids, or finding MRI changes, or showing the presence of ragged red fibers on muscle biopsy only help strengthen the case for a mitochondrial cause; they do not provide a final diagnosis. Only defining a biochemical abnormality of respiratory chain function or, better still, finding the mutated gene provides certainty (Chapters 2, 3 and 4).

Treatment of mitochondrial psychiatric diseases: Currently, no guidelines are available for treating psychiatric disease caused by mitochondrial dysfunction. It has been suggested that mitochondrial disorders respond poorly to psychotropic medications [13] and that this may indeed be a clue to this etiology. Depression or psychosis caused by mitochondrial disease should be treated according to current guidelines, but due to potential mitochondrial toxicity, certain drugs should be avoided.

Neuroleptic drugs have well-known side effects, particularly extrapyramidal toxicity. Drugs such as chlorpromazine and haloperidol have also been shown to inhibit respiratory chain function, particularly complex I activity, in laboratory studies [14]. Mood stabilizers such as sodium valproate should also be avoided, particularly in patients with *POLG* mutations, in whom this drug can precipitate acute hepatic necrosis.

Summary: Mitochondrial disease can be associated with the whole range of psychiatric disease including anxiety, major depression and psychosis, and these can be a presenting feature in patients with otherwise unrecognized systemic involvement. Awareness of the possibility that mitochondrial dysfunction is the cause is the crucial step to making the diagnosis and care with choice of treatment essential in order to avoid potential toxicities.

CASE REPORT

Case Report 16.1

A 36-year-old man who was in full employment presented in status epilepticus: according to his wife, he had had two weeks of general malaise, nausea, headache and progressive loss of hearing. MRI showed changes in the right temporal and occipital lobes. An initial diagnosis of encephalitis was later revised due to progressive proximal weakness and ataxia, and he was subsequently found to have the m.3243A>G mtDNA mutation. Two years later, follow-up examination showed that he had clear gait ataxia, evidence of a myopathy and peripheral neuropathy, diabetes and a cardiomyopathy: his epilepsy was controlled with carbamazepine.

He was readmitted two years after diagnosis because of rapidly worsening hearing and longer-term problems with memory, including episodes of confusion. During this admission, he became acutely psychotic with visual and auditory hallucinations, periods of aggression and increasing confusion. His electroencephalogram (EEG) showed encephalopathic changes (with widespread slow wave activity), but no obvious focal seizure activity, and his MRI showed new diffusion weighted changes in the left temporal lobe compatible with cytotoxic edema and a new "stroke-like" lesion. He was eventually transferred to the psychiatric hospital, where his symptoms were controlled using olanzapine.

His subsequent course was defined by gradual cognitive decline and sensory isolation. Hearing aids were initially helpful, but later failed to provide any functional hearing (he was not considered for cochlear implantation due to his poor cardiac and general condition). Initially, he had problems with aggression, but subsequently became increasingly apathetic, losing interest in family members and social contacts despite retaining reasonable physical health: his heart failure was well controlled and, apart from occasional seizures, so was his epilepsy. Within five years, he became fully dependent on help and required nursing home care. He died eight years after diagnosis.

Self-Assessment Questions

- Can you think of two potential causes of psychosis in this patient?
- Do you agree with the management approach with reference to his epileptic activity?
- Are olanzapine and haloperidol safe to use in patients with mitochondrial disease?

Answers

- The two potential causes of psychosis in this patient were (1) the hallucinations could have been driven by focal epileptic discharge. While we did not see evidence of seizure activity, this cannot be discounted. (2) Acute hearing impairment on a background of organic brain disease could also have precipitated a psychosis. While he had no earlier psychiatric illnesses, we should be sensitized to this possibility.
- Treatment of his epilepsy was crucial. We obtained a good response with carbamazepine and he remained well controlled for most of his life. In retrospect, however, we could have been more aggressive in treating the possibility that the psychotic manifestations could be epileptic in origin. He responded reasonably well to antipsychotic medication, but this case has changed my management, making me more inclined to start aggressive epilepsy treatment for this type of symptom.
- Olanzapine appeared to control his psychosis and was not associated with obvious side effects or worsening of his mitochondrial disease. He did receive haloperidol on a number of occasions, and this too did not seem to have any adverse effects, demonstrating that, in need, we can use these medications. Of course the theoretical concern of neuroleptic malignant syndrome in the context of someone with an already raised creatine kinase (CK) and metabolic impairment could result in significant metabolic disturbance and a prolonged period of rehabilitation not to mention a prolonged and complicated Intensive Therapy Unit (ITU) stay (see Chapter 12); however, acute psychosis must also be managed and cannot be suboptimally treated.

Approach

This man has the typical and full-blown syndrome associated with the m.3243A>G mutation that we often abbreviate and call the MELAS syndrome. His presentation with status epilepticus is not uncommon and he goes on to develop a multisystem disorder with ataxia, diabetes, cardiomyopathy, skeletal myopathy and peripheral neuropathy.

His psychiatric problems are most likely organic and related to his brain pathology. His initial presentation is precipitated by the lesions in his right temporal and, to a lesser degree, occipital lobes. These "ischemic-looking" lesions (which are acute energy deficits) are potentially affecting the auditory cortex and can explain his relatively acute hearing impairment (we did not test for comprehension, unfortunately). The second acute episode involved the contralateral temporal lobe, making the hearing disturbance complete. His subsequent course reflects the global cerebral and cerebellar neurodegeneration that is typical of MELAS.

Along with adequate and proactive management of his seizures and psychosis, long-term management of his hearing loss is also vital, as this can be a major reason for social isolation. It can, moreover, compound the problems of physical disability, making the patient even more at risk of developing psychiatric disease.

References

1. Inczedy-Farkas, G., et al., Psychiatric symptoms of patients with primary mitochondrial DNA disorders. *Behav Brain Funct* 2012; **8**: 9.

2. Gardner, A., et al., Alterations of mitochondrial function and correlations with personality traits in selected major depressive disorder patients. *J Affect Disord* 2003; **76**(1–3): 55–68.

3. Wang, G. X., et al., Mitochondrial haplogroups and hypervariable region polymorphisms in schizophrenia: A case-control study. *Psychiatry Res* 2013; **209**(3): 279–283.

4. Anglin, R. E., et al., The psychiatric manifestations of mitochondrial disorders: A case and review of the literature. *J Clin Psychiatry* 2012; **73**(4): 506–512.

5. Fattal, O., et al., Psychiatric comorbidity in 36 adults with mitochondrial cytopathies. *CNS Spectr* 2007; **12**(6): 429–438.

6. Anglin, R. E., et al., The psychiatric presentation of mitochondrial disorders in adults. *J Neuropsychiatry Clin Neurosci* 2012; **24**(4): 394–409.

7. Tzoulis, C. and L. A. Bindoff. Serial diffusion imaging in a case of mitochondrial encephalomyopathy, lactic acidosis, and stroke-like episodes. Stroke 2009; **40**(2): e 15–7.

8. Tzoulis, C. and L. A. Bindoff. Acute mitochondrial encephalopathy reflects neuronal energy failure irrespective of which genome the genetic defect affects. *Brain* 2012; **135**(Pt. 12): 3627–3634.

9. Boles, R. G., et al. Cyclical vomiting syndrome and mitochondrial DNA mutations. *Lancet* 1997; **350**: 1299–1300.

10. Chinnery, P. F. and D. M. Turnbull. Vomiting, anorexia, and mitochondrial DNA disease. *Lancet* 1998; **351** (9100): 448.

11. Bindoff, L. A., Mitochondrial gastroenterology, in *Mitochondrial Medicine*, S. DiMauro, M. Hirano and E. A. Schon, Editors. 2006, Informa Health Care.

12. Gramstad, A., et al., Neuropsychological performance in patients with POLG1 mutations and the syndrome of mitochondrial spinocerebellar ataxia and epilepsy (MSCAE). *Epilepsy Behav* 2009; **16** (1): 172–174.

13. Anglin, R., P. Rosebush and M. Mazurek Psychotropic medications and mitochondrial toxicity. *Nat Rev Neurosci* 2012; **13**(9): 650.

14. Burkhardt, C., et al., Neuroleptic medications inhibit complex-I of the electron-transport chain. *Annals of Neurology* 1993; **33**(5): 512–517.

Index